D1595717

Reflections on Lynn Margulis

"I greatly admire Lynn Margulis's sheer courage and stamina in sticking by the endosymbiosis theory, and carrying it through from being an unorthodoxy to an orthodoxy. I'm referring to the theory that the eukaryotic cell is a symbiotic union of primitive prokaryotic cells. This is one of the great achievements of twentieth-century evolutionary biology, and I greatly admire her for it."

—**RICHARD DAWKINS**, evolutionary biologist, and author of *The Selfish Gene* (from *The Third Culture* by John Brockman)

"Lynn Margulis was an independent, gifted, and spirited biologist who learned as early as the fourth grade to 'tell bullshit from . . . real authentic experience,' as she put it in a 2004 interview. With courage, intellect, a twinkle in her eyes, and considerable fortitude, she changed our view of cellular evolution."

—**DR. JAMES A. LAKE**, Distinguished Professor of Molecular, Cellular, and Developmental Biology and Human Genetics, UCLA, and winner of the 2011 Darwin Wallace Medal (from *Nature*)

"She was one of two or three great biologists of the twentieth century in most people's opinion. She inspired lots of people to go into the field. Even if people disagreed with her, she was very much a larger-than-life figure."

—**DR. NICK LANE**, evolutionary biochemist, University College of London Department of Genetics, Evolution, and Environment (from BBC Radio *Last Word*)

"Lynn Margulis was among the most creative challengers of mainstream Darwinian thinking of the late twentieth century. . . . Like other mavericks I have met, Margulis could not help but yearn, now and then, to be a respected member of the status quo, whose work merely confirmed the prevailing paradigm. But without courageous rebels like her, science would never achieve any progress."

—**JOHN HORGAN**, director, Center for Science Writings at Stevens Institute of Technology (from *Scientific American*)

"I admired her for many years from her writings and from hearsay. I was very fortunate to meet her two years ago. At a dinner party, I witnessed her defend the Gaia hypothesis against what another biologist present had said in print. She had the unfortunate person cornered; she was able to quote, word for word from memory, what he'd said, and she was very intent on having him see why it was wrong. I must say that when I witnessed this conversation I was reminded of the accounts written of Galileo when he came to Rome, in which he is described as defending the Copernican hypothesis at dinner parties in the houses of the great families there. I saw in her the same confidence in her vision, together with impatience at those who can't think as openly or as broadly but instead choose to misunderstand the new ideas. I've thought for many years that we as yet barely understand the implications of Darwin's discovery that we evolved via natural selection. I'm sure that Lynn Margulis has seen further than most what this means for our view of the natural world and our relationship to it."

—**LEE SMOLIN**, researcher, Perimeter Institute for Theoretical Physics, author of *The Life of the Cosmos*, *Three Roads to Quantum Gravity*, and *The Trouble with Physics* (from *The Third Culture* by John Brockman)

"I had the pleasure of knowing Lynn for most of my life. Both Dorion and I had parents involved in the Bostonian scientific community, and we became fast friends at an early age, so I also got to experience some of Lynn's style of mind-nurturing. She was awfully busy, and not always around, but occasional outings to places like science museums were thrilling. She spoke to kids as if they could understand adult concepts, which they generally can, and I still remember puzzling over something she said about metabolism when I was, oh, probably ten. Later I learned how actively she mentored scores of grad students. Now a vast army of her former students, many of whom also consider her an invaluable friend, populates the halls of biology and exobiology."

—**DR. DAVID GRINSPOON**, curator of astrobiology, Denver Museum of Nature & Science and chair of astrobiology, the Library of Congress (from *Pale Blue Blog*)

"Lynn could be pretty wild. Sometimes it was vexing, but I loved her for it. She revolutionized our view of the living world, pushing on one door after another, behind each of which flourishes, still, an alternate and, to some, unthinkable conception of biology and evolution. As it's turning out, many of these doors lead to astonishing realities, others to pure imagination, but each rewards any who enter with a great intellectual challenge. Whether chatting or cooking with Lynn, you would find yourself being led to and through one after another of these doors. It was great fun. I think there were many recent critics of her later hypotheses who clearly did not grasp the rules of the game. We will all miss her something terrific."

—**DR. LES KAUFMAN**, professor of biology, Boston University

"We have lost a great independent thinker; I wish more scientists were like her . . ."

—**DR. ROALD HOFFMANN**, Nobel laureate, Frank H. T. Rhodes Professor of Humane Letters Emeritus, Cornell University

"We might even say that it was Lynn Margulis, not Charles Darwin, who actually explained the mechanics of the origin of species."

—**DR. MARK MCMENAMIN**, Mount Holyoke College, Department of Geology and Geography (from *21st Century Science & Technology*)

". . . Basically, Lynn was way, way ahead of her time."

—**DR. MARGARET MCFALL-NGAI**, professor, Department of Medical Microbiology & Immunology, University of Wisconsin–Madison (from *On Wisconsin*)

Lynn Margulis

The Life and Legacy of
a **SCIENTIFIC REBEL**

EDITED BY **DORION SAGAN**

CHELSEA GREEN PUBLISHING
WHITE RIVER JUNCTION, VERMONT

A Sciencewriters Book

scientific knowledge through enchantment
Sciencewriters Books is an imprint of Chelsea Green Publishing. Founded and codirected by Lynn Margulis and Dorion Sagan, Sciencewriters is an educational partnership devoted to advancing science through enchantment in the form of the finest possible books, videos, and other media.

Project Manager: Patricia Stone
Developmental Editor: Brianne Goodspeed
Copy Editor: Susan Barnett
Proofreader: Alice Colwell
Indexer: Shana Milkie
Designer: Melissa Jacobson

Chelsea Green Publishing is committed to preserving ancient forests and natural resources. We elected to print this title on 30-percent postconsumer recycled paper, processed chlorine-free. As a result, for this printing, we have saved:

7 Trees (40' tall and 6-8" diameter)
3 Million BTUs of Total Energy
618 Pounds of Greenhouse Gases
3,351 Gallons of Wastewater
224 Pounds of Solid Waste

Chelsea Green Publishing made this paper choice because we and our printer, Thomson-Shore, Inc., are members of the Green Press Initiative, a nonprofit program dedicated to supporting authors, publishers, and suppliers in their efforts to reduce their use of fiber obtained from endangered forests. For more information, visit: www.greenpressinitiative.org.

Environmental impact estimates were made using the Environmental Defense Paper Calculator. For more information visit: www.papercalculator.org.

Part-opening: Dark-field photomicrograph of the planktonic foraminiferan *Globigerinoides sacculifer*, an open-ocean, single-celled organism by David A. Caron

Printed in the United States of America
First printing September, 2012
10 9 8 7 6 5 4 3 2 1 12 13 14 15 16

Our Commitment to Green Publishing
Chelsea Green sees publishing as a tool for cultural change and ecological stewardship. We strive to align our book manufacturing practices with our editorial mission and to reduce the impact of our business enterprise in the environment. We print our books and catalogs on chlorine-free recycled paper, using vegetable-based inks whenever possible. This book may cost slightly more because it was printed on paper that contains recycled fiber, and we hope you'll agree that it's worth it. Chelsea Green is a member of the Green Press Initiative (www.greenpressinitiative.org), a nonprofit coalition of publishers, manufacturers, and authors working to protect the world's endangered forests and conserve natural resources. *Lynn Margulis: The Life and Legacy of a Scientific Rebel* was printed on FSC®-certified paper supplied by Thomson-Shore that contains at least 30% postconsumer recycled fiber.

Library of Congress Cataloging-in-Publication Data is available upon request.

Chelsea Green Publishing
85 North Main Street, Suite 120
White River Junction, VT 05001
(802) 295-6300
www.chelseagreen.com

I don't consider my ideas controversial.
I consider them right.

—Lynn Margulis interview with Dick Teresi

Contents

REBEL, TEACHER, NEIGHBOR, FRIEND

INTRODUCTION

Indomitable Lynn

DORION SAGAN

Unlike many, perhaps, she continued to grow and learn until the end. One of the last projects she was involved in was the characterization of the symbiotic bryozoan *Pectinatella magnifica*, who, like Lynn, loved to dwell in the possibility of Puffers Pond, the lake across which she swam nearly every day that last blue summer of 2011. There it was she quoted to me the words of Emily Dickinson, "That it will never come again/Is what makes life so sweet."

Harsh and true and beautiful—like her. But, defying Dickinson, I'll throw my boat in with Anaïs Nin, who said that we write to taste life twice, in the moment and in retrospection. Thinking of those gentle summer ripples, that blue pond in whose water the matter of my mother's body now lies, I am reminded of Krug, a fictional character in Nabokov's *Bend Sinister*, who as the novelist describes it, "in a sudden moonburst of madness, understands that he is in good hands: nothing on earth really matters, there is nothing to fear, and death is but a question of style, a mere literary device, a musical resolution." If she burned out in a sudden burst of hemorrhagic overactivity, like a blazing celestial object vanishing into its own glory, the end-blaze was not so different than the burning life, as she died near the height of her powers, at the peak of her coruscating personality.

Death is a mystery, but it is also a sublation, a rising up to another realm, a reckoning, not just a negation but a planting and possible flowering, an archiving, a setting of the seed of the soul as information into the fertile if not eternal field of collective memory. Of course memory, as every novelist and memoir writer knows, cohabits with and is infused with imagination, prey to the twin temptations of revisionism and hagiography. The dear disappeared, no longer able to speak on her own behalf, to mischievously interrupt, to make her own case and sweet diatribes, is made to mean what we want her to mean, instead of saying what she would have said, perpetuating her own brand of perplexity. In the case of Lynn, my matrix, my secular miracle of a mother and writing partner, that perplexing brand consisted of several scientific strands, woven into a rope of such strength one imagines it suitable for performing the Indian rope trick or, with the fibers differently arrayed, into that fabric of lore permitting lucky riders upon a flying carpet. Her science took her, and us with it, along for a once-in-a-lifetime ride not only into previously unrevealed regions within the wild geography of intellectual inquiry, but also back into the time of Earth's earliest beginnings, when, as she never tired of pointing out, so very much had happened. Growing the grass for us even as we are paving the way—or maybe not—for a future we naïvely imagine is obligated to include us.

When I used to stay at her house in Amherst, I slept downstairs in a drawing room with two sets of creaky sliding doors, one wall lined with books. Holding his ground on a shelf in the northeast corner, watching over me in my sleep, was her pre-Columbian man, a beloved piece of statuary that (I found this part out after she died) she picked up for $500 after a tiff with my stepdad about his losing the same amount in a poker game, and which she willed upon her death returned to Mexico, its country of origin. When I awoke I would be treated to five portraits, of my mother and her sisters as little girls. There were two of her. In one she's a baby, not yet two. In the other she is maybe six or seven. It awakes in me a curious affection, the same I experienced at seeing them while she was still alive, bustling about with inimitable energy at the age of seventy-two, running to feed and warm, in of all places, and not without an unappetizing odor, the microwave, dog food for her black Irish wolfhound, the rescued and eternally grateful Menina (named after the Velázquez painting), as she prepared to meet Jim MacAllister to swim, before preparing for classes, for lectures, for guests coming or going, for children and grandchildren, papers, books, and projects without end. Food for her was fuel, and her energy such that toddlers had been known to request naps after being drained by the continuousness of her curiosity, the boundlessness of her enterprise. She grew up in the South Side of Chicago during the Depression, and this experience, along with her poverty as a young mother and woman, fostered in her a lifelong appreciation of culinary frugality, of leftovers. She once was caught eating cereal with slices of hot dog in it and explained that the dog (a different one, the scrappy mutt Roosevelt) didn't want it.

She had many motivators, from the desire to get out of the constant bickering between her parents, which led her into the beautiful act of rebellion that was getting into the University of Chicago at the age of fourteen, to the startling death of her mother, Leone, who dropped dead one day in her late middle age after exiting the shower. Her mother's stroke imposed upon Lynn a sense of urgency to finish her many and ever-expanding circle of scientific projects centered around the unveiling of early life and characterizing the evolutionary and ecological nature of life on earth. This boundless energy that ran toddlers ragged and could exasperate as well as entice, educate, and enchant those around her—a motley mix of students, children, colleagues, and friends swept up into the whirlwind that was Lynn—made it right to describe her, as Earthwatch founder Brian Rosborough did, as a life force.

In some of her effects in my possession, I have a copy of a letter from my father to an esteemed academic colleague, who is duly informed of my father's progress and plans in all things scientific and astronomical. If all goes well, Carl concludes, he will receive his PhD in approximately nine months. I was curious, seeing the date of the letter, March 18, 1959, to see if there would be any mention of family in this unstoppable passion for science that also, not uncoincidentally, infected his young wife and former teenage girlfriend, my equally if not more unstoppable mother. And yes, there it was—almost a footnote in this letter detailing lunar organic syntheses from Yerkes Observatory to Dr. H. J. Muller in Bloomington, Indiana—"Speaking of nine months, Lynn gave birth yesterday to an 8 lb. 2 oz. Boy, Dorion Solomon Sagan. It feels strange adding our fiber to the red thread. I've never before had so strong a feeling of being a transitional creature, at some vague intermediary position between the primeval mud and the stars." That was me, although I'll have to check my birth certificate because I always thought it was 8.6 pounds, but right or wrong one can glean the passion my parents had at that time, succored by America's post-*Sputnik* gravy train, for the life of science, for the exploration of new worlds that my dad, in that same letter, compares to the age of exploration as Europeans began to venture across the seas at the beginning of the fifteenth century.

In Carl's letters as in Lynn's life, the scientific and the personal are mixed, as are the private and the public, as the great gulf stream of the discoverer's quest sweeps up all—family, friends, lovers, and lives—into its world historical quest. This was the life of my father, and my mother caught its fire and sent it in a different direction: instead of imagining icy floaters in the ammonia atmosphere of Jupiter, or underneath the regolith of Mars, she delved into that mud that my father had taken more for granted in his extraterrestrial flights of fancy, and explored the real organisms—the methanogens and archaeans, the symbiotic protists and wriggling corkscrew-shaped spirochetes—beneath her hiking boots. She did so with gusto and love, and once she got started she never stopped. In the volume before you, you will see her in her journey, her stubborn recalcitrance, her passionate quests, her fearless interactions with nature inclusive of stuffed shirts, old boys' networks, and patriarchal power structures that would intimidate a lesser mortal. For her, it was the opposite: she often intimidated them. She was too often right, too articulate and passionate, too well versed in the minutiae of chemistry, ecology, evolutionary theory, cell biology, microbiology, geology, and a thorough, encyclopedic familiarity with the ultimate objects (and subjects) of natural history, organisms themselves.

In this book you will sample her indomitable personality as she sampled the multicolored microbial mats, the sulfurous seaside mats and inland muds, the waterlogged logs and forest floors from which she derived her photosynthetic bacteria, her spirochetes, her symbionts, her wood-eating termites, sometimes smuggled into the country with the cellulose-digesting microbial communities in their hindguts fully intact. You will read of her interactions with those fellow life-forms, also deeply if often unknowingly symbiotic, whom she derided as a species but appreciated as individuals. You will hear of how she tweaked, countered, incensed, and ultimately often intellectually triumphed over her contemporary great evolutionary biologists. The Spanish biologist Mónica Solé Rojo told me, for example, of Stephen Jay Gould, inarguably in the top percentile of articulateness, dumbfounded in the presence of her onslaught of supporting examples to her argument, replete with species names. Much the same, indeed more, can be said and herein will be, of the friction between her and the Oxford zoologist Richard Dawkins, whose popular "selfish gene" concept in her opinion missed so much in terms of subtlety and reality with regard to the real living world, symbiotic and transformative, that she knew so well.

Despite the rancor she elicited as a result of her forthright and fearless manner, and the adamancy with which she insisted on her points of view, the more so when, though supported with crucial facts they were underdogs in the temporarily prevailing but ultimately malleable scientific consensus, Lynn also stood out because she could change her mind. She could learn, and continued to do so, to the end of her days. It is true that at times she may have been attracted to unorthodox views because they needed an ally in a world of consensus genuflectors and pig-pilers, a world, paradoxically, where her own views on symbiotic interliving were deemed untenable by coteries of coactors teaming up and cooperating against a perceived intruder like an immune system detecting a foreign protein or a quorum of bacteria sensing, as one being, a threat. But her threat was not to people but to the evil done to the spirit by the entrenchment of unsupported views.

Although she could be a bulldog, her heart was soft and her spirit loving beneath the scientific realpolitik of her conversation and the insistent tough-mindedness of her sometimes strident and blunt, withering and refreshingly unadorned opinions. That deep intellectual honesty didn't always need to be right in details to be right in approach, which it was, and could not else but be for her. In a privileged place to observe her intellectual and personal trajectory, I became more enamored of her as she grew older and I tagged along, always a lucky twenty-one years behind her rushed schedule. At the end of her life I became more fond of her as a person, an individual wonderful in her own right and separate from the matrix that spawned and coddled me, cuddled and argued with me, gave me simple gifts of clothes and books and peasant food, who always had demands and expectations and, as might be surmised, had trouble simply relaxing. Time was too short, life was too important to squander it on trivial pursuits.

But as she grew older there were more signs of an increasing capacity for being happy. Her laugh was deeper, the twinkle in her green eyes more mischievous. She was less likely to be rankled, quicker to accept the latest instances of perceived incompetence, inaccuracy, and hubris from the benighted race into which, through an amorous accident and hundreds of millions of years of microbial evolution, she had curiously been born. As I looked at those pictures in the drawing room where I slept on the fold-out futon she had replaced with a new one to add to my comfort at my home away from home, which was and always will be my home, I noticed in myself a feeling of fatherly tenderness for that little girl whose endless curiosity stared out in perfect composure from the pictures on the wall. At seven years of age, there she was, an inchoate nestling and intimation of future greatness, a force of nature in seedling form, as yet to be loosed, not least by herself, onto a not-always-appreciative world. I had read Roland Barthes's essay on photographs of his mother engendering a similar feeling of reversed paternity and was thus predisposed.

When she became seventy-two, and reiterated a refrain from a sparse but multiyear discourse on the bewilderment of change and aging and her reluctance to make an unnecessary to-do about it, I commented to her that I had recently come across a study that said the age of maximum subjectively identified happiness was seventy-three. Since you are seventy-two, I joked, at least you'll have something to look forward to. As chance or fate or Gaia or some other agglomeration of that ambivalent nexus of causation the Greeks explored in tragic theater would have it, she did in fact die at the age of seventy-three. Pictures of her, for example her beaming visage over the podium in a poster that advertised the last lecture of hers I attended—to a packed house of UMass students as she detailed her views of life on earth and her discovery of a symbiotic representative, the brain-shaped, log-clutching bryozoan *Pectinatella magnifica*, at hers and their local swimming hole, Puffers Pond—showed that increasing capacity for happiness as she came closer to the end of that trajectory that catapulted her from the land of the living, of flesh and blood, to the murkier, ecological realm whose inhabitants and chemical cycles she herself helped characterize.

My son (for whose care as a boy she provided, giving invaluable love and assistance) Tonio and I kayaked to the center of Puffers Pond at the end of a private family ceremony. There we sprinkled from a pink urn of Himalayan salt my son's grandmother's ashes back into the water, inhabited by the bryozoan she discovered, the water across which she swam nearly every day that last summer of her life, the summer of 2011, her dog sniffing and running about on the shore. Tonio dropped a coin of remembrance into the waters (in olden days to pay Charon to ferry one's soul safely across to the world below) and her ashes spread, after lingering, as with ghostly mischief and microbial mystery, accommodating the subtle ripples and disturbance from the paddles and sinking coin. As I tried to explain to all nine of her grandchildren, including two of her beautiful granddaughters (who bear an uncanny resemblance to the pictures that used to grace the far walls of that room whose futon she'd replaced), Lynn's body is gone, she has returned to the nature she so loved and studied, but part of her is still here, left behind as they themselves, their hearts and thoughts and smiling, curious faces.

Beginnings

Tale of Tales

JORGE WAGENSBERG

It may be the fable of fables, the tale of tales: Once upon a time, a long, long, *long* time ago, only bacteria lived on our planet.

Even with nobody else around, these bacteria had a problem: those that ate well lumbered slowly under their bulk, while those who moved swift and limber through the waters, they went—wriggled is more like it—hungry.

It wasn't all that bad, though. The mutual limitations of the big slow ones and the quick hungry ones kept everyone from eating one another. This fortuitous deprivation regulated competition, allowing both populations to persist without using up all available resources.

So far so good—or at least not all bad.

Then one day, a big, fat bacterium ate a skinny, squirmy bacterium that was not well fed but was a fast and excellent swimmer. Actually, this didn't happen just once. It happened billions of times during life's infancy on our planet.

On this particular day something different and truly extraordinary happened. The fat bacterium didn't digest the athletic one. Instead, they merged. Without speaking—in the ancient language of biochemistry, a text of molecules and textures rather than words—Big Bacterium put forth a symbiotic proposition. He wanted to form the ultimate, the perfect, total, symbiotic pact: a merger not of possessions, or of lips or other body parts, but of whole selves.

"Why don't we move the way you do," said Big Bacterium, "and eat the way I know how?"

And thus it was that Fast Bacterium no longer had to go hungry, but rather was cared for within the translucent bounty of greater semipermeable beings. And thus it was that Big Bacterium, although he had done quite well for himself on his own, now had a means of propulsion—as did his offspring, who were as multifarious as the sea was wide or the sky high.

Truly, it was a pact most fantastic. It was as if a rock had found wings to fly, or a snail had invented motorized wheels.

In the tale of tales, it happened overnight, and everyone lived prolifically ever after. In fact, however, this symbiotic pact was not easy. It took an astonishing 1.5 billion years from the first of these proposed symbioses—that of wriggling spirochetes combining with archaea—to evolve a cell with a nucleus, a cell that looks like a paramecium or amoeba.

This fable, of course, is that of Lynn Margulis, biology's greatest heroine. It illustrates the birth of the first eukaryotic cell, the cell that created your entire living world—all animals, plants, and fungi—by the deep and permanent merging of different kinds of bacteria. Although the earliest part of it, the blending of speedy spirochetes with larger bacterial cells, has not been by any means universally accepted, the second and third parts of the fable, which involve the merging of primitive beings to make the first animal-like cells, have been. It is now agreed, based on genetic evidence, that the oxygen-using parts of cells, the mitochondria, and the photosynthetic parts of the cells of plants and algae, came about when our ancient ancestors, archaea and bacteria, merged. The fabulous mode of separate beings coming together, blending as one, is how we got here.

And the lesson could not be more clear: collaboration can outcompete the competition.

The idea of symbiogenesis—the coming together of different organisms to make viable, distinct new ones—is biology's most beautiful and powerful idea after natural selection.

Although Lynn Margulis deserves credit for making the world recognize its truth, the idea was not original to her. As often occurs, the idea that triumphs is not well understood by the scientific community of its time. Other evolutionary thinkers, such as the German Andreas

Schimper (1883), the Russian Konstantin Merezhkovsky (1909), the Frenchman Paul Portier (1918), and the American Ivan Wallin (1927), had all speculated similarly while facing the incomprehension and withering criticism of their colleagues.

A young Lynn Margulis fell in love with symbiosis—it was she who finally managed to precisely describe the stages in the process that lead bacteria to become eukaryotic cells—and was emboldened rather than dissuaded by criticism. It was by no means easy. Her seminal article, "On the Origin of Mitosing Cells," was published in 1967 in the bold and prestigious *Journal of Theoretical Biology*, but only after being rejected fifteen times (fifteen times!) by other first-class publications. But thanks to her insistence, this fable—technically, SET, or serial endosymbiosis theory—became accepted as true.

Jorge Wagensberg teaches theory of irreversible processes in the faculty of physics at the University of Barcelona. One of Spain's most prominent communicators of science, he was director of CosmoCaixa, the science museum in Barcelona that was awarded the European Museum of the Year Award in 2006.

Erudition

MOSELIO SCHAECHTER

There are two kinds of great scientists: scientists known for their impressive experiments, and scientists who make groundbreaking theoretical syntheses. Lynn Margulis was an example of the latter. She is responsible for the transformative idea that eukaryotic cells (from yeasts to vertebrates) evolved by the acquisition and exploitation of other, smaller cells, a process known as endosymbiosis. Accordingly, essential components of eukaryotic cells—the organelles mitochondria and, in photosynthetic cells, plastids—are derived from bacteria that some ancestral cell had ingested. These events are thought to have taken place early in the history of life on earth.

In 1967, in a fifty-page-long article in the *Journal of Theoretical Biology*, Margulis, then only twenty-nine, presented this thesis, resting it on a mountain of data. Her arguments referred to neglected publications, and she gave credit to those who had proposed the same notion. Despite its antecedents and extensive documentation in her paper, it took some ten years after her publication for the hypothesis to become one of the central tenets of modern biology. Ultimately she was recognized by being elected to the National Academy of Sciences and receiving, in 1999, from US President Bill Clinton, the National Medal of Science.

Margulis's detailed theory led to predictions. Margulis correctly predicted: "If an organelle originated as a free-living cell, it is possible that naturally occurring counterparts still can be found among extant organisms."

It is now widely accepted that relatives of bacterial ancestors of mitochondria and plastids exist to this day. In support of this idea, both mitochondria and plastids have critical biochemical commonalities with bacteria. But the clincher lies in their genomic similarity: they share genes with free-living bacteria. These organelles have streamlined genomes, sometimes consisting of a few dozen genes, yet their genetics testifies to a common heritage. The mitochondria in your cells are obvious genetic cousins of *Rickettsiae*—which, interestingly, exist only as intracellular parasites—while genetic studies trace plastids, the photosynthetic parts of plant cells, to photosynthetic cyanobacteria.

The importance of this notion extends beyond explaining steps early in evolution because it also explains what came later. It empowers the momentous idea that evolution does not proceed by single mutational steps only but, notably, also occurs by the acquisition of packets of genes simultaneously. The allure here is that it allows us to grasp how complex structures may have arisen not one mutational step at a time, but by acquiring and combining multiple arrays of genes that themselves had evolved for different functions. Evolution not just as a slow and random slog, but an evolution that hops, skips, and jumps as organisms move toward, with, and into one another.

In Margulis's weltanschauung, symbiosis was *the* driving force of evolution. Note, however, that symbiosis is not the only way to acquire new genes en masse. One of the other powerful discoveries in recent times has been that bacteria acquire such packets via the incorporation of viruses or plasmids—a view that Margulis viewed with some reticence. I never asked her about this, but she could have claimed that, by extension, these entities could also qualify as symbionts. She proposed that evolution by symbiosis, or "symbiogenesis," could explain many specific phenomena. Notably, she championed the idea that eukaryotes gained their cilia (which differ from the prokaryotic flagella—and which she insisted be called "undulipodia") by the acquisition of a motile bacterium, a spirochete, a view that has few subscribers today. She also proposed that AIDS is caused by a long-living (ergo symbiont-like) spirochete, an even less popular view. But those who actually read her seminal 1967 paper should have seen such idiosyncratic ideas coming: it is written with the fierce intensity that became her trademark.

Her positive contributions extend well beyond proposing scientific theories. She wrote a number of books that had a profound effect on readers. Many young people have recalled their pleasure in reading them and have noted, not infrequently, that it influenced their decision to pursue a career in science. Her familiarity with the inner lives of members of certain biological groups was stunning. Much was acquired the old-fashioned way, by going on field trips and collecting material for the lab. She was a master of the protists—organisms that are not bacteria, plants, animals, or fungi. Through her knowledge and insight, she made others aware of the importance of this highly diverse group of organisms from which plants, fungi, and animals evolved.

Margulis's depth of knowledge of protists included a critical understanding of the pioneering work done in the early twentieth century and before. Such erudition is manifest in the 1967 paper: it was not just a manifesto on the endosymbiotic theory; it was also a treatise on comparative biochemistry, cytology, and systematics.

I met Lynn long ago. Perhaps around 1970, Lynn and I frequently went on walks with our little ones. We lived in Newton, Massachusetts, perhaps a mile apart, and our spouses were glad to have some time to themselves on Sunday mornings. Weather permitting, we always went to the same place, somewhere in Needham, near the Charles River. My memories are of stimulating conversations, which is self-evident considering Lynn, but I can't recall specifics. I don't think we talked much about science, at least not in detail. I have a (much less developed) natural history bone in my body, so there was always something like that to share. We developed a strong friendship. Although we saw each other only occasionally, our encounters were inevitably joyful. We had similar, or at least compatible, interests and attitudes. On occasion, the notions she espoused caused dismay among her colleagues and friends. But dismay is not her legacy. She should be remembered as a person whose passionate beliefs and creative ideas, some of them quite iconoclastic, changed scientific discourse in several areas, affecting both scientists and nonscientists.

Moselio Schaechter is former president of the American Society of Microbiology. His writings can be found at Small Things Considered: The Microbe Blog. *His essay here previously appeared in a slightly different form in* Science.

As Above, So Below

ANDRE KHALIL

Lynn Margulis, biologist and Distinguished Professor of Geosciences, composed a grand and powerful view of the living and the nonliving. Integrating the work of obscure Russian scientists, DNA pulled from cell organelles, computer-generated daisies, and the hindguts of termites, her vision was wider in scope and more profound in depth than any other coherent scientific worldview. At the time of her death on November 22, 2011, it was a vision that remained misunderstood and misconstrued by many scientists.

Much of this view came from her uncanny ability to first lean forward and see the smallest inhabitants of the earth; to hover there, and then to leap back at the speed of thought to conceptualize the entire planet. Lean forward, then stand back. This inner movement, this seeing from soil to space, marked a unique scientific endeavor.

This perspective was earned only through walking through diverse areas of study—geology, genetics, biology, chemistry, literature, embryology, paleontology. Those fields are sometimes separated by an untraversed distance at universities: they are housed in separate buildings that may as well be different worlds. In Margulis, they found agreement and discussion with each other; they were reconnected, just as they are intrinsically connected in nature.

This journey led her to emphasize in all her scientific work two phenomena—the fusing of distinct beings into a single being: symbiosis;

and the interaction of organisms and their environments to create relational loops that led to regulation of many earth systems: Gaia theory.

Taken separately these concepts have the ability to redefine, respectively, how we understand organisms and the environment.

Taken together, they can redefine our consciousness.

After the earth was born, give or take a few hundred million years, there were bacteria. Bacteria were here first and are with us still, comprising a major part of the biosphere. They are unseen with the naked eye; they lack nuclei (for this reason, they are called prokaryotes—"pro" = before, "karyon" = nucleus). Their forms were legion, and their metabolisms were, and continue to be, strange.

Where life could exist, it did exist in these tiny forms. One of these forms was most likely an amorphous blob like *Thermoplasma.* This bacterium was quite fond of heat and sulfur, the stuff we now associate with the devil. Another bacterium was the spirochete. Familiar to us now as the type of bacterium that causes syphilis and Lyme disease, the spirochete is a curl of an organism, a tremulous and crooked line with no front or back. Margulis studied these strange beings through literature and microscope. From some corner of her intellect, they called to her.

The thermoplasmid and spirochete of early Earth were neighbors and, in a sense, enemies. Each one would try, when it encountered the other, to consume it. This was a popular notion at the time: meet and consume. Soon enough, encounter after encounter between the two beings led to an unprecedented event: the beings came together to eat each other and decided on marriage instead. Just what changes happened to cause this friendly ingestion is still unknown. What is known is that the spirochete didn't digest the thermoplasmid and the thermoplasmid did not digest the spirochete. As Margulis was fond of saying, "1 + 1 = 1." There was a union of the two, resulting in an entirely new being. They were inseparable, literally. The thermoplasmid had a rotor now, and the spirochete had a "head." A head and a tail: for the first time, beings had direction. Cultural philosopher William Irwin Thompson examines this emergence in his book *Coming into Being.* It isn't that spirochetes couldn't pursue a coordinate before—but the asymmetricality of the new, combined entity resulted in a new way of being, completely without reference in the history of life: one end, distinct in form, ingested the food; the other end did the rowing. Both absorbed the nutrition. This was a giant step in the evolution of consciousness and is echoed by all true evolutions in consciousness: the rise of a new way of being, inconceivable to the world that came before.

And soon other mergers were taking place. Oxygen-breathing bacteria were incorporated by endosymbiosis into this being. Where once oxygen was poison, now it flowed through without harm.

Cyanobacteria, green and photosynthetic, were incorporated in some of these cells as well. Both these symbioses remain visible today—as the mitochondria in all cells (the oxygen-breathing bacteria that became mitochondria) and chloroplasts in plant and some animal cells (the cyanobacteria that led to chloroplasts). These are ancient partnerships that have never dissolved and that continue to pulse with rhythm. Our existence depends upon them. Human cells reflect these unions, and we breathe plant-respired oxygen.

Margulis, inspired by the work of little-known biologists, revealed and proved these mergers for us. At first her work was rejected and scoffed at. It did not fit the still dominant neo-Darwinian paradigm that tells us all evolutionary novelty comes from natural selection acting on genes and the gradual accumulation of random genetic mutation. But eventually these symbioses were accepted because they could not be ignored. In a stunning display of reluctance, despite mounting evidence,[1] the spirochetal origin of the undulipodium (sometimes incorrectly called or mistaken for the flagellum, though the undulipodium and flagellum are not similar either chemically or structurally) is still contested and sometimes dismissed.

What is unquestionable: bacteria make up the living architecture of our bodies. They evolved into our cells and also remain "free-living" in our digestive system. Their spiraling remnants are in our guts, our brains. This means our physical selves are universes composed of the movements, biological agreements, and interactions of these beings.

What can this mean for the individual? What happens when we are simultaneously songs and compositions of notes? "Identity is not an object; it is a process with addresses for all the different directions and dimensions in which it moves,"[2] Margulis once stated, with her colleague Ricardo Guerrero.

What happens when we are notes, songs, and the notes again? What happens when we shift our perspective and see that we are cells made out of cells?

As above, so below and as below, so above. Margulis, somewhere in the middle, decided to thoughtfully occupy both positions. "Why does everybody agree that atmospheric oxygen . . . comes from life, but no one speaks about the other atmospheric gases coming from life?" she asked. Bacteria created a whole other host of these gases, as Margulis knew well from her work. After she found James Lovelock, they worked on making those processes known. Their collaboration resulted in Gaia theory, which was a disciplinary symbiosis—the theoretical expression of Margulis's interdisciplinary life.

Gaia is the work of the relational loops of push and pull between bacteria, other organisms, and the environment. The clouds, the atmospheric gases, the pH and salinity of the ocean, and other earth systems express the "dialogue" between the organisms and the earth. This dialogue is Gaia theory. Particularly relevant to these relational (often called "feedback") loops are the smallest living beings, the bacteria. In this dialogue, the information yielded from and received by the bacteria and environment is crucial to the existence of life on this planet. Remove the bacteria and everything dies. The world becomes a Mars or a Venus, overtaken by harshness or billowing clouds so thick that everything is obscured. No direction-creating spirochetes and *Thermoplasma*; no respiring green cyanobacteria; no purpose or breath; and there is no biosphere, for bacteria are its regulators.

The science behind Gaia, particularly that found in Lovelock's formulations, is complex and detailed, not guesswork. Lovelock came up with an understandable and accessible metaphor in the form of a computer program called Daisyworld. Daisyworld is not the "proof" of Gaia: Lovelock and his colleague Andrew Watson devised the program to see if living and environmental factors could theoretically interact without intention. This was a rebuff to the many criticisms that Gaia had to act through some sort of New Age benevolence. This view might be acceptable in spiritual circles but is damning in scientific ones, and so: Lovelock's little model.

In Daisyworld there are black daisies, which absorb the sun's heat, and white daisies, which reflect heat. Both flowers grow and produce offspring, and both have the same thresholds for life and growth—they cannot grow at a low temperature and die at too high a temperature. The black daisies, which absorb heat, grow faster in cooler conditions; since the heat accumulates in their petals. White daisies, which reflect the heat, need warmer conditions to produce more offspring and thrive. The sun that shines on Daisyworld is dynamic. It grows in luminosity over millions of years.

Here is Margulis, quoted at length from *Symbiotic Planet*, to make clear the results.

> Without any extraneous assumptions, without sex or evolution, without mystical presuppositions of planetary consciousness, the daisies of Daisyworld cool their world despite their warming sun. As the sun increases in luminosity, the black daisies grow, expanding their surface area, absorbing heat, and heating up their surroundings. As the black daisies heat up more of the surrounding land surface, the surface itself warms, permitting even more population growth. The positive feedback continues until daisy growth has so heated the surroundings that white daisies began to crowd out the black ones. Being less absorbent and more reflective, the white daisies begin to cool down the planet. . . . Despite the ever-hotter sun, the planet maintains a long plateau of stable temperatures.

Many additional factors have been added into subsequent Daisyworld models. The little world has always displayed a deep relationship between species selection and planetary temperature regulation.

The environment could no longer be seen as a tyrant, lording over selection; it was now a co-evolving field. And all the organisms on the planet are connected by this vast system of regulation and dynamism. "Gaia," Margulis's former student Greg Hinkle said, "is just symbiosis as seen from space."

Nothing 20 kilometers up or down on the earth escapes the pulse of collectivity. Indeed, no action or process is untouched by it, even the action of evolution itself.

Margulis's answer to evolution was a logical extension of her work: evolution happens through symbioses and Gaia.

That symbiosis caused evolutionary innovation was readily observable in microorganisms, in large part because of Margulis's work. But neo-Darwinists such as Richard Dawkins still refused to accept it as true in the case of multicellular organisms, and thus tried and continue to try to discredit the theory.

Unfortunately for them, it's not so simple: Gaia processes are real and observable (and sometimes referred to as "biogeochemistry," a term more acceptable to mainstream science). Furthermore, the five kingdoms (bacteria, protoctists, fungi, plants, animals) of life are all touched by symbiosis. The bacteria are the symbionts. The protoctists (mistakenly called "protozoa"; they are not animals, so the "zoo" in the word is a misnomer) readily display symbioses. Indeed, symbiogenesis has been observed in the lab. An amoeba population accidentally infected with bacteria was observed over long periods of time, and soon enough the infecting bacteria could not be removed from the infected amoeba without killing the organism.

Since 99.9 percent of all organisms on the planet are microbial beings, if we're talking about evolution, we must be talking about microbes. Richard Dawkins himself admitted as much in a debate with Margulis at Oxford, when he said he could not claim to know much about life, since he knew very little about bacteria. Animals, plants, and fungi readily display symbiotic mergers as well. It's not just that all eukaryotic (nucleated) cells are the products of symbiosis. All animals have symbiotic partners in their guts. Remove these symbionts and the animals die. Because of the disparity in size, we have trouble thinking of a rabbit as a symbiont with bacteria, but it is.

Margulis and her son and coauthor, Dorion Sagan, present it this way in their book of strange, otherworldly brilliance, *Acquiring Genomes:* "Darwin's question about how species originate may be rephrased as: 'What is passed from parent to descendant that we detect as evolutionary novelty?' A straightforward answer is, 'Populations and communities of microbes.'"

I call the book's brilliance "strange" and "otherworldly" because in it, a new view of the world rises to the surface. *Acquiring Genomes,* along with another of Margulis and Sagan's books *Microcosmos*, show us a bacterial view of the world. Bacteria exchange their genes laterally. This means they don't pass their genomic information only when they reproduce, though this can happen, but also through their simple existence. Bits of their genomes float in and out of their bodies and into other bacteria. This was, and is, happening all the time. The web of life created by such gene transfers is unbelievably complex and can even be baffling.

Along with the many detailed examples of bacterial mergers at varying levels of cellular complexity, the world revealed by *Acquiring Genomes* is also a world of mating between distinct phyla (a classification just below "kingdom"—meaning creatures of different phyla vary wildly from one another). This phenomenon, which should not be possible according to scientific orthodoxy, has been shown by UK scientist Don Williamson. Again, Margulis's work has been contested, but she and Williamson have collected vast amounts of data and evidence, including live examples demonstrated in physician and writer Frank Ryan's *The Mystery of Metamorphosis*. Many people dismissed Margulis for this large-scale sexual symbiosis, through which genomes are transmitted from one totally different being to another; but most of them have not looked deeply into Williamson's work, and certainly not his live samples, preferring instead to dismiss without real investigation. Margulis was working on this project at the time of her death, and it remains to be seen whether or not other scientists will champion Williamson. Like much of Margulis's work, it requires the uncommon ability to question basic assumptions to even understand the phenomena.

All her efforts and ideas, those accepted and those still controversial, led Margulis to sharply criticize the standard neo-Darwinist theory of evolution.

It's not that she didn't understand it, as some of her critics liked to claim. Margulis has examined natural selection and genetic mutation carefully. In fact, Gaia theory is an intense examination of natural selection, since Gaia's processes of regulation are the "natural selectors." The push and pull of the biota (the total sum of all organisms) and the inorganic—their weaving and separations, their gestures of relationship—set the framework of regulation. There is no need to be vague about "fitness" and just what the environment "selects" with Gaia in the picture. Instead, there is something to aim for—exploring Gaia's processes of regulations.

But Gaia, the natural selector, does not create from the top down alone. While natural selection can refine all beings, no new species have been shown to arise from the natural selection plus random genetic mutation model. The difference between refinement and speciation is one that confounds and also confuses neo-Darwinists, who cart out example after example of refinement as proof of their theories, not realizing that they are still not indicating true speciation. Darwin himself did this by using dog breeding as evidence for his theory. Alfred Russel Wallace, who codiscovered evolution and whose view differs from Darwin's in significant ways, referred to this as "unnatural selection" and was keen to note that it could not represent real evolutionary change.

Symbiogenesis may not prove to be the beginning and end of evolution. After all, it does not explain why forms are expressed in the way that they are (e.g., why should similar gene sets express themselves in one creature as feathers and in another as spores?). These laws of nature remain to be revealed, but have been pursued in innovative ways by thinkers as disparate in time and field as Johann Wolfgang von Goethe (who thought of certain forms as having a "blueprint" of archetypal reality that bloomed into specific forms) and Brian Goodwin (who looked at evolution as a movement of physical and mathematical laws). What is definite is that the merging of beings is key, and symbiogenesis offers a clearly observable alternative to the consistent but woefully incomplete neo-Darwinian paradigm.

The neo-Darwinists were equally critical of Margulis's work, some going so far as to say she was "corrupted by fame"—presumably the slight fame she achieved after she popularized the endosymbiotic origin of cell organelles. Anyone who knew Margulis laughed at such accusations. She worked in a small lab with a few dedicated graduate students. The lab was small in part because she resisted funding from corporate and governmental agencies that she thought would damage the integrity of her work. Once she dismissed a potential funder for wanting her to do work whose content could not be disclosed to the public. "If it's not public, it's not science," she said, and hung up the phone on tens of thousands, possibly millions, of dollars. The graduate students were dedicated because she practiced science for science's sake, and was fond of quoting quantum physicist and philosopher David Bohm, who said, "Science is the search for truth . . . whether we like it or not." The truth was Margulis's concern, not popularity, not big money, and certainly not fame.

Many neo-Darwinist concerns circled nervously around words like "Gaia" and "cooperation" (which Margulis did not like to use). They were, perhaps rightly, concerned that these terms were ripe for religious appropriation. Margulis herself was outspoken against such mishandling of her research.

Some New Agers love to grasp symbiosis as signifying altruism between organisms. It's much more complex than that—there is something "in it" for every symbiont, just as a state beneficial in some way arises out of each symbiosis. Terms like "altruism" have no scientific value, because they are too single-minded to describe the phenomenon.

New Age thinkers also use "Gaia" as a blanket term. They've appropriated it to mean that the earth is a living organism. Or they refer to Gaia as a "goddess." This turns Gaia into a sort of Stepford planet by containing its complexity in a simple and inadequate metaphor. This no more grasps reality than "selfishness" does our genes.

Margulis expressed her solution to the error once by saying, "Gaia is not merely an organism." The earth is beyond stale conception. It is more magnificent and active than we can imagine. Gaia is object and process. Gaia houses volcanoes and every book, every word on volcanoes ever written, and at the same time is those volcanoes. It is where our greatest loves live, and where every human heartbeat has ever rhythmically pulsed. In this new understanding, that something can pulse with life and yet be beyond our concepts of living, those concepts begin to change.

If Gaia is conscious, it possesses a consciousness of a different magnitude, probably of a different order altogether.

Richard Dawkins and his precursors, such as John Maynard Smith, as well as other misguided neo-Darwinist thinkers, could not and cannot understand this lesson: this complexity is impossible to incorporate in a linear and reductive understanding.

Part of their failure lies in a misunderstood version of cause and effect that plagues science. At a certain level of complexity, somewhere just above a billiard ball clanking into another billiard ball, cause and effect begin to change shape. This change may be real: that is, it may actually shift in its laws and patterns in nature. Or it may be imagined: in other words, it may demand a different sort of thinking. Effectively it doesn't matter, since we need to contend with the shift in our thinking. To encompass complex systems with our thinking, we must imagine a model that is less like "cause-effect" and more like "being-manifestation." That is, multiple layers and numerous agents of forces unconsciously conspire together, and their conspiring is so intermingled that it is simultaneously cause and effect, and thus beyond both. For example, the being, or process, of Gaia manifests itself as an unstable, constantly correcting level of oceanic salinity. One cannot be said to cause the other, since the oceanic salinity interacts so deeply with the beings and environs from which it arises. Symbiosis and biological forms demand the same sort of thought.

This complexity shames the metaphorical lack of nuance in "selfish genes." Neo-Darwinists, who so often speak publicly about the erosion of sound scientific thought, have themselves engendered ideas that represent a threat to clear scientific thinking. It's not merely that Dawkins's metaphors are incorrect (and they are incorrect), but his whole idea of evolution is too mystical (in the pejorative sense), too imagined, too metaphorical to be correct. Dawkins, who claims to be an atheist, relies on a host of selfish angels within us and the possibility for meme salvation to justify his theory. He substantiates his magical worldview on a meager past of scientific work.

Margulis, on the other hand, worked constantly and tirelessly in her lab, always aiming at and incorporating new pursuits. At the time of her death, she—with her handful of graduate students and a clutch of international scientists as collaborators—was researching cures for Lyme disease and reassessing how treatable syphilis is (both Lyme and syphilis come from spirochetes, which Margulis probably knew more about than any other scientist); she was also writing a book on Emily Dickinson. Her projects often had the unsettling side effect of forcing us to reexamine our most cherished presumptions. In other words, she was a sort of investigative light where Dawkins is merely polemical shadow: she was a true materialist whose work produced spiritual effects.

Neo-Darwinism is an evolution that people can and have built social theories (memes, for example) out of. But symbiogenesis and Gaia theory, truer versions of evolutionary motivators, require a new philosophy and perspective to understand at all.

It requires the deepening of the capacity to understand.

These concepts are not conveniently, like neo-Darwinism, mirror images of the current economic system (nor are they, as many confusedly think, a Kropotkinan "mutual aid" analogue for socialism) and so have enjoyed no real social metaphor. Perhaps as we—in the newly and deeply connected world of the Internet, social profiles, and globalization—witness the dissolution of the cult of isolated individuality and embark on understanding a clearer and more nuanced view of individuality, so too will we ready ourselves for a clearer view of evolution and life.

"In the arithmetic of life, One is always Many. Many often make one, and one, when looked at more closely, can be seen to be composed of many,"[3] said Margulis and Guerrero. Being able to move from one perspectival state to the next—this is a sort of mental-phase transition that is necessary to understand life, evolution, and the environment. It is the sort of thinking Goethe advocated, a thinking whose movement mirrored the movement of life itself.

Margulis grasped this before us. She did more than any other scientist in recent history to expand and explain this. Presented in the essay is only a small sample of her visionary works. It isn't always easy to grasp her thinking, nor to rise to the challenges of it. It is much easier to dismiss complexity and reduce ourselves to smaller ideas. Now that Margulis has died, it remains our choice to catch up with what she and her life's work have set in motion. To do so, we must bring together the many fields of knowledge she embodied. Biologists must talk to physicists, virologists must talk to geologists, cosmologists must talk to microbiologists, and scientists must talk to nonscientists. This motion of meeting and exchanging ideas, if we act with it, will evolve our thinking.

Andre Khalil is also known as Conner Habib, a writer, adult film performer, and teacher. He has written and presented widely on the topics of sex, science, and cultural criticism. He studied organismic and evolutionary biology in graduate school under Lynn Margulis and now lives in San Francisco, where he runs a Rudolf Steiner discussion group.

On Lynn
from a Close Friend and Colleague
JAMES LOVELOCK

Today, November 24, 2011, is Thanksgiving, and I'm still shaken by the news that Lynn died two days ago. The news came from her lab by that laconic and heartless medium, email. I come from an older generation and still expect deeply moving or important news to come by a personal letter, not as a message sandwiched between slices of spam. It will take a long time to digest the fact that Lynn is no longer with us. The idea that Earth is a live planet that regulates its surface and atmosphere in the interests of the biosphere is intimately connected with us both. It arose in my mind at the Jet Propulsion Laboratory in 1965, when I shared an office in the Space Science building with Carl Sagan. At that time it was no more than an unusual idea that I shared with a small group of space, atmospheric, and climate scientists and three years before my friend the Nobel Prize–winning novelist Bill Golding gave the concept the name Gaia in homage to that classical Greek goddess of the earth. Lynn, at the time, was developing her theory of endosymbiosis, which is closely linked to Gaia, but it was seven years before we met as colleagues in 1972.

She had written inviting me to call at her lab when next I was in the United States to talk about oxygen in the atmosphere, so I traveled to Boston from New Hampshire by bus, met Lynn at the airport, and traveled with her on the MTA to her lab at Boston University. She was the first biologist I had met who became excited by the idea of Gaia, and certainly the first to take it seriously. After a lively afternoon talking with her and her students, I asked if she could recommend a nearby motel where I could stay so that we might continue the discussion next day. Lynn said, "No, why not come home with me and meet my family? We can continue the discussion there and you can travel back here with me tomorrow."

It was not long before a full collaboration was in progress, and it led to the publication of two papers on Gaia, one for the Swedish journal *Tellus* and the other for *Biosciences*. In some ways it was more like a revolutionary war than a collaboration. Lynn was like a wartime general who led her troops from the front; she went into combat against the cronies of the earth and life sciences firmly established in their turf dugouts. The war went on for nearly thirty years, until a partial peace was declared in 2001 by the Amsterdam declaration, where it was agreed that the earth was indeed a self-regulating system comprising all forms of life, the air, the ocean, and the crustal rocks. That peace was reluctantly signed by university-based scientists who wanted to keep their own sciences intact and not share them with those who occupied the other buildings around the campus. It was well put by the eminent geologist Dick Holland, of Harvard University, who referred to Gaia as "a charming idea, but not needed to explain the facts of the Earth." Not surprisingly Gaia is still not favored by established science. The battle goes on, and we will all sorely miss Lynn's fearless and forthright outbursts. I recall her reply to an ill-informed scientist who made the claim that one or other of the alleged threats to the environment would destroy all life on earth: "Gaia is a tough bitch." For that audience, it was just right.

Battling scientists usually fight with sharpened pens dipped in acid ink or with words spoken through subtly distorting megaphones, not with atom bombs. Lynn's style of fighting, one that would have met with the approval of her fellow American General Patton, did not go uncriticized. Lynn and I often argued, as good collaborators should, and we wrangled over the intricate finer points of self-regulation, but always remained good friends, perhaps because we were confident that we were right. In these times of Facebook and letting it all hang out, many might find it hard to understand how we could work closely together and yet not be romantically involved. The nearest we came to intimacy was in 1972, when I had my first heart attack in the road just outside and then inside Lynn's home in Newton, a suburb of Boston.

I might not have written this had not Lynn and her husband, Nicky, given the help and support I needed.

It is interesting to me that our battles with other scientists were limited to those in the earth and life science departments of universities. Physicists and chemists were properly neutral, and climate scientists and meteorologists often welcomed Gaia. We collaborated with several scientists at that cathedral of science, the National Center for Atmospheric Research (NCAR) at Boulder, Colorado, from 1962 until the 1990s. They included James Lodge, Will Kellogg, Steve Schneider, Lee Klinger, and Robert Dickinson. During the 1980s, there was almost a censorship by peer reviewers of any paper about Gaia, unless it was critical of it.

Apart from the distinguished geologist Robert Garrels, geologists were, like Dick Holland, quietly dismissive of Gaia and remained so until the 1990s. But an increasing number of earth scientists came out of the sediments and began to realize that the earth did indeed regulate its climate and chemistry. Disliking the name Gaia, with its New Age associations, they called their neogeology "earth system science."

Evolutionary biologists, especially neo-Darwinists, were among Lynn's favorite targets and soon the arguments became so fierce that at one point the talented wordsmith and neo-Darwinist Richard Dawkins referred to Lynn as "Attila the hen" and the distinguished English neo-Darwinist John Maynard Smith referred to Gaia as an evil religion. When Ford Doolittle published his now famous critique of Gaia in *CoEvolution Quarterly*, I could not stand aside and let this well-written and apparently logical demolition of Gaia become the last word. After some fairly ineffectual attempts to compose a verbal response, it occurred to me that so complex were the factors determining the mechanism of a planetary self-regulating system, that a properly mathematical computer model was needed as an answer. It is important to know that even the simplest of self-regulating mechanisms resists rational explanation; to explain them requires a circular argument. Cause-and-effect thinking, so prevalent in science, fails to explain the facts of physiology, quantum physics, and many other real but inexplicable dynamic systems.

I took time off and composed a computer program for the mathematical model, which is now known as Daisyworld. I launched it at a meeting hosted by our good friend Peter Westbroek in 1978 on the Dutch island of Walcheren. I knew that Daisyworld was a definitive answer to the neo-Darwinists' criticisms of Gaia and that it must be properly published in a peer-reviewed journal. This was done in collaboration with another friend, Andrew Watson, who among other qualities is a competent mathematician. So unpopular was Gaia then, that despite my track record of numerous previous papers published in *Nature*, the journal would not take our Daisyworld paper. It did not matter too much because the highly regarded Swedish journal *Tellus* took it.

The emergence of Daisyworld marked a watershed in the development of Gaia as a theory of the earth. A great deal of the further papers on Gaia were about mathematical models that descended naturally from Daisyworld. They involved extensive collaborations with my colleagues Tim Lenton and Stefan Harding. At the same time it led to a drifting apart of the collaboration that Lynn and I had together. We remained close friends but returned to our original scientific bases, biology for Lynn and transdisciplinary science for me.

For me, and I hope eventually for most of science, Lynn's greatest contributions were in cellular biology. She discovered endosymbiosis, the process by which the complex eukaryotic cells of present-day life evolved through the successful fusion of simpler and singular prokaryotes, bacteria. This is a key step in the evolution of life on earth. Her great contribution to Gaia was to show that microorganisms now, and from the beginning, were the infrastructure of Gaia. Our tendency to ignore bacteria is an example of our false pride. Lynn was the first to tell me that we humans are mere cellular communities. They are huge ones, comprising 10 billion living cells, but 90 percent of these are not human cells but cells of other microorganisms, most of which evolved to be friendly.

The history of Gaia might be summarized by saying that I had, thanks to NASA, a top-down view of the earth through telescopes and spectrometers, and saw it as a system where the biosphere regulated its climate and chemistry. This was in 1965 and when astronauts, and all of you vicariously, saw directly through your eyes that blue/white iconic sphere. In the early 1970s, Lynn gave us all the bottom-up view of Gaia through her microscope and showed that it was made of microorganisms and alive.

James Lovelock, inventor, chemist, and originator of the Gaia hypothesis, is author of several best-selling books, including The Vanishing Face of Gaia: A Final Warning.

Gaia Is Not an Organism:

Scenes from the Early Scientific Collaboration between Lynn Margulis and James Lovelock

BRUCE CLARKE

One of the most fortunate collaborations of modern science took root in 1971, near the outset of Lynn Margulis's professional career, prior to her receiving tenure at Boston University and well before her later fame and notoriety. For readers more familiar with her later writings, her earliest papers—both on microbial evolution and, coauthored with James Lovelock, on Gaia—show at times an unexpected mixture of theoretical daring and professional hedging. In the early 1970s, despite her radical proposals already extant for rethinking the origins of eukaryotic cells, Margulis is still feeling her way beyond mainstream neo-Darwinism and into her mature constructions of symbiogenesis as a major evolutionary dynamic. From this angle, her early correspondence and publications with Lovelock do not present scenes of seamless confirmation of mutual prior convictions. Rather, we see them co-laboring over an arduous crucible wherein, throughout the 1970s, both her science and her scientific persona are developed and fully forged.

But first, in order to set up the historical and intellectual stakes of that discussion, I will take a moment to tell my own conversion story. Doing so will bring out a view of "autopoietic Gaia"—the divergent form Gaia theory takes in Margulis's later presentations. This is also the form in which Gaia finally got through to me. Since that moment, I have been a vocal proponent of Gaia theory—a critical and selective disciple, I hope. But for a long time I was reluctant to take it seriously. Meeting Lynn Margulis in the fall of 2005, then in the fall of 2006 spending two weeks at her lab after attending a Gaia theory conference at George Mason University in her company, clinched my change of mind. But my epiphany proper, the moment of intellectual conversion from vague skeptic to Gaian thinker, occurred, after a long preparation, just before those first personal contacts, tripped by an encounter with her memoir, *Symbiotic Planet*.

Even after more than a decade of cultivating a post-tenure academic specialization in literature and science, as a professor in a department of English reschooling myself in the histories of physics, chemistry, and biology and coming up to speed on chaos and complexity theory, throughout that period in the 1990s, where Gaia was concerned, not much came to hand. I absorbed the usual nebulous and unexamined notion that Gaia was not quite real science but some New Age notion connected to goddess worship or God knows what. I took it to be the sort of idea that I, as an interloper into the discourse of the sciences, were I to establish or maintain some minimal credibility in the academy, should avoid—just as one should avoid contact with bacteria, since, you know, they're nothing but germs and they'll make you sick.

At some point around the turn of the millennium, looking for an accessible introduction to biology to teach in my undergraduate literature and science class at Texas Tech University, I was browsing the science shelves at Lubbock's Barnes & Noble bookstore. I was hoping to find something more recent but in the vein of Lewis Thomas's delightful *The Lives of a Cell*. What I found was the paperback of the popular-science text *What Is Life?* by Lynn Margulis and Dorion Sagan. I recollected then that Margulis's science in the early 1970s had been the source of many of Thomas's best zingers. Here was that same exciting science, expanded, updated, and set forth in equally if not more glorious expository prose. Yes—this will do just fine. I began teaching *What Is Life?* flanked from semester to semester by hard bioscience fiction by authors such as Ursula Le Guin, Bruce Sterling, Octavia Butler, Paul Di Filippo, and Joan Slonczewski.

What Is Life? gives a brief but straightforward introduction to Gaia theory. This was probably my first encounter with an authoritative account of it. However, it does not bring Gaia theory forward so emphatically that one must confront it head-on. I taught this text for several years, concentrating on its main evolutionary narrative—a theme still conducive of existential tremors in a good number of my West Texas undergraduates—while otherwise sweeping Gaia off into a neglected corner. My turning point was catalyzed elsewhere. It came in the train of developing a parallel interest in Niklas Luhmann's sociological systems theory, which features a dramatic extension of the concept of autopoiesis beyond its origin in biological systems theory. Luhmann's work follows a line of systems theory termed second-order cybernetics, in which autopoiesis is a primary if not the premier concept. For Humberto Maturana and Francisco Varela—the Chilean biologists who forged that concept—their initial, material instance of an autopoietic system is the living cell.

In this formulation, the fundamental processes of life are circular in form—they continuously select and transform the elements in their environmental medium so as to produce their own continuing production of selective transformations. For the theory of autopoiesis, cellular life's self-referential processes amount to cognition, or sense-making. In *What Is Life?* Margulis and Sagan call this "sentience." Fatefully, Margulis and Sagan are perhaps the only biological science writers other than Maturana and Varela to have incorporated the concept of autopoiesis into popular expositions. So, inadvertently working on these two dialects of systems theory simultaneously, I gradually saw that there were links to be made between, on the one hand, the second-order cybernetic discourses of autopoiesis and, on the other, the evolutionary and symbiogenetic applications of the concept presented in Margulis and Sagan.

I decided to give my next literature and science class a biocybernetic orientation and teach Maturana and Varela's *The Tree of Knowledge.* Instead of assigning the relatively lengthy and rigorous *What Is Life?* for a biology primer, I went with Margulis's briefer, single-authored text of 1998, *Symbiotic Planet.* Preparing it in the summer of 2005, I got around to its final chapter, simply titled "Gaia." Here she retells the name-of-Gaia origin story, but with a cautionary twist:

> The term *Gaia* was suggested to Lovelock by the novelist William Golding, author of *Lord of the Flies*. In the early 1970s, they both lived in Bowerchalke, Wiltshire, England. Lovelock asked his neighbor whether he could replace the cumbersome phrase "a cybernetic system with homeostatic tendencies as detected by chemical anomalies in the Earth's atmosphere" with a term meaning "Earth." "I need a good four-letter word," he said. On walks around the countryside in that gorgeous part of southern England near the chalk downs, Golding suggested Gaia. . . . The name caught on all too well.[1]

Margulis expresses her concern that "Gaia"—with its extrascientific baggage, along with Lovelock's propensity in popular discussion to personify Gaia as a living being—has exposed the science it covers to severe misconstructions. But then comes, for me, the key part of this passage, the piece that finally made the idea of Gaia click. Margulis continues: "As detailed in Jim's theory about the planetary system, Gaia is not an organism" (119). This blunt proposition cuts away from the understanding of Gaia the fringe metaphysics or planetary vitalisms kept alive, so to speak, by the name of Gaia itself having "caught on all too well." For her own part, in this passage Margulis points to the scientific details of Lovelock's developed presentation of the theory in order to tether the metaphorics back to the science of Gaia. It is not an organism; rather, Gaia is a system, a metabiotic system within which organisms are elements. Having put the organic metaphor in its proper place, Margulis goes on to deploy it: "Gaia, the system, emerges from ten million or more connected living species that form its incessantly active body" (119).

This subtle but incisive separation of rhetoric from exposition was my eureka moment: Gaia theory *is* systems theory. And not only that: in her own later treatments, Gaia theory is second-order systems theory. Margulis has prepared Gaia for coordination with the suite of autopoietic systems theories also making their paradigm-changing way against institutional and ideological headwinds. Specifically, on a par with psychic and social systems in Luhmann's adaptation of autopoiesis, Gaia is a metabiotic entity, a complex emergent system with its own evolving metadynamics, arising from the inextricable interpenetration of the biota with the seas, the skies, and the rocks. Now I was ready to go back to *What Is Life?* and construct Margulis and Sagan's phrasings there in their second-order cybernetic sense: "The biosphere as a whole is autopoietic in the sense that it maintains itself" (20). As an autopoietic system in the metabiotic sense, that is, Gaia need not be identified with the form of life per se. Rather, it participates in an essential quality of individual living systems—the autopoietic form of organization, an emergent, recursive form of self-production based on self-referential, self-regulated self-maintenance that crosses over between biotic and metabiotic application.

In telling my conversion story, I have also sketched some more recent chapters in the history of Gaia theory, in which the main emphases of Margulis and Lovelock diverge to some degree. I will turn back now to the chronology of Margulis's first involvement with it. The details of this part of the history are not particularly well known. For instance, an otherwise excellent and veracious obituary of Margulis posted on the Web site of the British newspaper *The Telegraph* commits several misstatements on this particular topic.

> It was Lynn Margulis's expertise in microbes that led her, in the mid-Seventies, to the British atmospheric chemist James Lovelock, who had come to suspect that living organisms had a greater effect on the atmosphere than was commonly recognized. Together they proposed a theory that Earth itself—its atmosphere, the geology and the organisms that inhabit it—is a self-regulating system in which living organisms help to regulate the terrestrial and atmospheric conditions that make the planet habitable.[2]

While this is an unusually up-to-date description of current Gaia theory, several of the historical statements here warrant correction. In their first papers Lovelock and Margulis present the Gaia *hypothesis*, a different, prior form of the theory. Margulis begins to collaborate with Lovelock in 1971, not in the mid-1970s. Lovelock has stated that, due to Carl Sagan's recommendation, Margulis initiated their correspondence "in the summer of 1970 . . . to ask about atmospheric oxygen."[3] However, Lovelock's first recognizably Gaian papers appear in the mid- to late 1960s. Lovelock debuts the name of Gaia before any coauthored

Lovelock and Margulis papers are in print. Lovelock defines Gaia at that moment as "a biological cybernetic system able to homeostat the planet for an optimum physical and chemical state appropriate to its current biosphere."[4] This is the proper form of the Gaia hypothesis at the outset of their collaboration. The hypothesis hinges on what was then the utterly heretical notion that *life controls the environment*: Gaia is "a *biological* cybernetic system." The chemist Lovelock had independently arrived at this biocybernetic orientation. In the development of the theory, he and Margulis will modify and temper this claim by redrawing the boundaries of the Gaian system to contain both the biota and the environment.

Additional historical assistance comes from Lovelock's original correspondence in the Margulis papers.[5] Interspersed there with a preponderance of his letters to her are occasional copies of her letters to him, as well as related documents such as rejection letters and readers' reports. Lovelock's first letter in the Margulis files is dated September 11, 1970. It does not mention "Gaia," only the scientific idea behind its initial formulation, delimited specifically to the atmospheric envelope: "I am in the course of writing a paper on the earth's atmosphere as a biological cybernetic system." He thanks her for sending him materials on the early evolution of cells and affirms his growing conviction that the primary components of the earth's atmosphere are "biologically maintained"—in other words, are produced and regulated in their proportions by living processes. By the spring of 1971 Margulis is reading Lovelock's essays and manuscripts related to the as yet unpublished Gaia hypothesis. These earliest papers address the interrelations of atmospheres and life. An open issue at the time is whether and to what extent the earth's atmosphere is biogenic. On March 31, 1971, a letter of hers addressed "Dear Dr. Lovelock" appears to document that Margulis did not yet fully grasp Lovelock's approach:

> Several of your charts are fascinating and very comprehensible (Table 1, Table 2 of planetary atmospheres) but where do these estimates come from? I'd really like to learn your methods for making these sorts of estimates as well as your sources of original data. Microbes strongly interact (i.e. take up, give off) hydrogen, nitrogen, ammonia, methane, carbon dioxide, oxygen, hydrogen sulfide at least and we are just beginning to

> know enough about the bugs in which these reactions occur to order them in an early-to-late evolutionary sequence. But how the gases themselves act in the environment . . . I really don't know what I need to learn or where to begin.[6]

Throughout that year, Lovelock's letters are similarly addressed, "Dear Dr. Margulis." In a letter dated September 17, 1971, he again declares the biocentric bent of his current theorizing: "The evidence in favor of the atmosphere as a biological contrivance grows." He confesses how difficult this is for him to write up, and how much he would "welcome a chance to exchange views." He praises her recent article, "Symbiosis and Evolution," in *Scientific American.* He then asks whether she would consider coauthoring a paper "on the atmosphere." The proposal for a formal collaboration comes from Lovelock and may be dated to this moment. A letter of his dated September 27 thanks her for considering while not yet agreeing to his offer of collaboration until after more conversation. They first meet in person when Lovelock comes through Boston around Christmas of that year. On January 3, 1972, Lovelock addresses his letter "Dear Lynn." He thanks her for the warm welcome and "stimulating even tho disjointed discussions," then mentions in passing that with regard to the idea of a "living planet[,] Bill Golding suggests 'Gaia' as a name for it." This timing may be why, despite the fact that Lovelock's confab with Bill Golding took place in 1967, in *Symbiotic Planet* Margulis states that Golding made that suggestion "in the early 1970s."[7] In fact, that would be the general date of the moment that Lovelock first mentioned "Gaia" to her.

It may be that, just to be on the safe side, he did not speak the name of Gaia until after he had secured her agreement to be his coauthor. That decision appears to have been the immediate scientific outcome of the social success of their first face-to-face meeting. On January 13, 1972, Lovelock writes to confirm his sense that they are now ready "for a clear cut paper with plausible supporting evidences." Margulis must have already been at work on it, as four days later he writes to thank her for sending a second draft. His previous letter covered the copy of a just-submitted manuscript that will appear later that year in *Atmospheric Environment.*[8] On January 24, Margulis writes back to comment on it. Her manner manifests her professional drive and her stringent editorial bent. Her remarks also indicate that she has now fully grasped Lovelock's implicitly Gaian argument:

> The mail will probably cross again. Anyway, I have read your oxygen article five times and finally not only do I dig it but I find it brilliant. Have you sent it in? Even so, I think you should strongly consider reducing the size, outlining the argument (I'll do this if you want my collaboration) and submitting it as a technical comment to Science in response to Lee [Leigh] Van Valens article. Van Valen raised this issue well but did not perceive the solution. I would also change the wording in several places to make it more transparent to the potential ecological and general biological audience. Please let me know soon what you think of this possibility. . . .

As far as I know, Lovelock did not act on Margulis's suggestion to retool this article and resubmit elsewhere. But the extant published paper, received in final form on February 21, 1972, does include Van Valen's article in the discussion.

On February 1, Margulis indicates further progress on their mutual manuscript. Around this time Lovelock decides to submit a letter to the editors of *Atmospheric Environment* as a supplement to the "oxygen article." It will go public with "Gaia" for the name of a hypothetical entity "with the powerful capacity to homeostat the planetary environment."[9] An undated letter from Lovelock written some time before February 16 discusses his decision. He expresses misgivings about it if Margulis feels that it will steal the thunder from their coauthored article in progress. He encourages her to coauthor and cosign this professional communication as well, especially given that its content has been "modified by our exchanges." Lovelock goes on in this long and crucial letter to Margulis to worry an issue that remains intrinsic to Gaian science, one of its key problems: how precisely is one to negotiate the boundary and distinction between living and nonliving systems? The letter in *Atmospheric Environment* states it this way:

> As yet there exists no formal statement of life from which an exclusive test could be designed to prove the presence of

> "Gaia" as a living entity. Fortunately such rigor is not usually expected in biology. . . . At present most biologists can be convinced that a creature is alive by arguments drawn from phenomenological evidence. The persistent ability to maintain a constant temperature and a compatible chemical composition in an environment which is changing or is perturbed if shown by a biological system would usually be accepted as evidence that it was alive. Let us consider the evidence of this nature which would point to the existence of Gaia. (579)

On February 16, 1972, Margulis addresses Lovelock's concern over the distribution of credit. Her own unconcern may be related to her already seeing the bigger picture and their roles in it as scientific revolutionaries:

> Jim, As for Gaia I do not in any way think you are preempting our mutual paper. On the contrary the more already in print and justified the better off we are. After all we are involved in attitudinal (scientific paradigm, Kuhn) change. Furthermore I really have not done the methane argument for myself in the detail I would like to before signing on. Go ahead and get it out on your own.

Lovelock sends off his single-authored Gaia letter to *Atmospheric Environment*. Around July they are ready to submit their first coauthored Gaia paper: "When the enclosed has received both your additions and your blessings I think it will be ready for *Science* (and when rejected minor modifications may fix it up for *Nature*)."[10] On September 22, 1972, the editor of *Science*, Philip H. Abelson, sends a rejection letter along with three readers' reports to Margulis on the submission of "The Earth's Atmosphere: Circulatory System of the Biosphere?" The biological emphasis of its title and the editor's referring to it as "your paper" argue that Margulis is its lead author. A month later Lovelock writes her: "Their editorial board must be senile."[11] It may be, however, that Margulis informed Lovelock of *Science*'s verdict without sending him the reports—two short notes, one detailed reading—for they are not hostile or closed-minded. Their gist is that to adequately pursue its thesis the paper needs more work.

When pieced together with some of Margulis's earliest single- and coauthored articles, this documentation shines a new light on the formation of her scientific commitments in the realm of evolutionary theory. A particular piece of the detailed reader's report stands out for its critique of the "optimization" claim, a line of cybernetic argument formulated by Lovelock during the early days of the Gaia hypothesis. This reader notes that, "As a basis for optimization neo-Darwinian or selective process is referred to," and then goes on to offer counter-arguments against that thesis. I speculate that we learn something unexpected from this: their earliest coauthored Gaia manuscript ventured a "modern synthesis" of Lovelock's optimization claim with a line of mainstream evolutionary argument. Now, in the 1980s and beyond, it is Lovelock who labors to have Gaia theory conform to Darwinian expectations, while Margulis is moving into the post-Darwinian realms of autopoietic organizations and laterally acquired genomes in the higher phyla. But both of these developments occur only after Richard Dawkins's and Ford Doolittle's neo-Darwinist critiques of Lovelock's first book, *Gaia: A New Look at Life on Earth,* published in 1979. Rather, in these very early years of the Gaia hypothesis, I think it is Margulis who is worrying the evolutionary issue as that emerges from her particular specialization. If I am correct in my chronologies, then at the outset of their Gaia collaboration, we find Margulis making the sorts of rhetorical appeals that she will later reject altogether. This also suggests that the evolutionary paradigm for which she is now best known was forged in part through prolonged efforts to articulate Gaian science.

A similar doctrinal appeal to neo-Darwinism shows up again just a year later in one of their first coauthored Gaia publications. The situation is like that with a Beatles song: they are all signed "Lennon-McCartney," but when Paul sings it, it is mostly Paul who wrote it. Despite Lovelock's name appearing as lead author of "Homeostatic Tendencies of the Earth's Atmosphere," there is good reason to think that Margulis took the lead in this composition as well. A letter of Lovelock's dated March 14, 1973, covers the draft of a paper he will present a month later in Mainz, with publication to follow in the journal *Tellus*.[12] He notes how the enclosed draft is "slanted from a Phys Chem outlook," whereas a different coauthored article in the works tells "the Biological version of the story." Lovelock pleads here, "Lynn I really do think and wish that you would be first author on the Exobiology chapter. . . . Don't let your kindness and consideration cause me to appear as grasping and egomanic in this Gaia adventure." The "Exobiology chapter" would be the "Homeostatic Tendencies" paper, published in *Origins of Life*, a journal then closely connected to NASA's Exobiology program. In it we read: "Although the environmental control mechanisms are likely to be subtle and complex, we believe their evolution can be comprehended broadly in terms of Neo-darwinian thought (Mayr, 1972)."[13] I take the parenthetical citation to acclaimed evolutionary biologist Ernst Mayr to pin the evocation of neo-Darwinism to an authoritative name. Such incidental arguments from authority mark an uncharacteristic tentativeness in these earliest Margulis-led Gaia papers.

One notes a comparably old-Darwinian rhetorical ploy in her earliest important semipopular article, "Symbiosis and Evolution," which narrates the classical Margulis thesis about the symbiotic origin of eukaryotic organelles while plying a progressivist rhetoric of evolutionary "advance" and "perfection," a rhetoric entirely expunged—indeed, powerfully argued against—in her mature work. What we learn from these examples, it seems to me, is that at first Margulis does not yet fully grasp the extent to which her evolutionary innovations challenge neo-Darwinian doctrines. Instead, she briefly ventures to prop up her radical proposals, singly and with Lovelock as coauthor, with some mainstream thinking. This does not last long: just as in the development of Gaia theory, Lovelock relinquishes the optimization argument, in time Margulis sloughs the neo-Darwinian veneer from her evolutionary arguments, Gaian and otherwise, in favor of what would now more aptly be called a systems-biological orientation. The intensity of her well-known antipathy to neo-Darwinian doctrines may have some roots in a personal reaction against these earlier professional compromises.

Occasional missteps aside, the abiding contribution that Lynn Margulis brings to the Gaia hypothesis in the first years of her collaboration with Lovelock is the addition of deep time, evolutionary depth. Lovelock concentrates on systems available for current inspection: "The best arguments in favor of Gaia come from the contemporary scene and it is on these I am concentrating. History is a mess."[14] He jokes that she prefers "probing around in a 3 billion year old septic tank."[15] Her own Gaian bent is gloriously messy, and it is fully in evidence in the first coauthored paper published with her as lead author, "Biological Modulation of the Earth's Atmosphere."[16] Its tables show histories of atmospheric gases and temperatures charted on time lines beginning with the origin of Earth, let alone the origin of life, followed by her trademark treatment of microbial evolution: "We emphasize the microbial contribution for two reasons: their metabolic versatility leading to profound environmental effects and because the regulation of the planetary environment was apparently proceeding long before the evolution of the larger (eukaryotic) life forms" (476).

We see that from the moment of its first publication, the Gaia concept spills over the notional boundaries of traditional scientific disciplines. In particular, both geology and biology had supposed a neat separation between the abiotic and the biotic realms. However, as natural processes tinker with their own evolving elements, breeding all manner of emergent formations and reality-testing their stability or viability—producing life out of nonlife, then more out of less complex life, then boot-strapping life and nonlife together into a metasystem—they do so without concern for the nice human constructions of peer-reviewed disciplinary distinctions. Even though these disciplines have existed in modern form only since the nineteenth century, they were codified before the rise of the systems sciences. Lovelock's conceptual breakthrough comes from cybernetics, the scientific metadiscipline that studies both artificial and natural "contrivances" for systemic operation and self-regulation. His early letters mentor Margulis on this: "once you have a homeostasis it is difficult to distinguish parameters from variables, inputs from outputs etc etc."[17] He puts Margulis on the winding road to autopoietic Gaia, a destination beyond both neo-Darwinism and Lovelock's own orientation toward control systems. Their collaboration brought together two great scientific outliers and interconnected and mutually strengthened their individual challenges to received disciplinary ideas. It has taken no less than the marriage of symbiogenesis to atmospheric chemistry to get Gaia taken as seriously as it is.

Bruce Clarke is the Paul Whitfield Horn Professor of Literature and Science at Texas Tech University.

Putting the Life Back into Biology

The Passionate Lynn Margulis

NILES ELDREDGE

What a way to meet someone: a few decades ago I was invited to speak at Amherst College. The details of what I had to say that evening escape me now—but I know it was about evolution and almost undoubtedly about species, speciation, and the fossil record. During the question period after the talk, someone in the audience asked me if anyone had managed experimentally to produce a new, fully reproductively isolated species in the lab.

I replied that Theodosius Dobzhansky had said that he had at first thought that someone (whether he himself, one of his students in his lab, or someone elsewhere—again I forget) had in fact done so with experimental populations of a species of *Drosophila*—but that it had turned out that it was not an instance of true speciation after all: the effect of reproductive isolation had been merely the consequence of infection of some of the fruit flies by some microbial organism or other. So, no, I said, no one had managed to produce a convincing, true case of reproductively isolated populations in the lab.

The explosion was instantaneous. A woman stood up midway near the aisle and started, well, shouting. At first I couldn't understand anything she was saying, but as she warmed to her topic—making her way to the aisle and stepping closer to the stage—I suddenly realized that this must be Lynn Margulis. *The* Lynn Margulis, already long known to me and the rest of the biological community as the framer and vocal proponent of the symbiotic origin of eukaryotes.

And I began to hear—and to listen to—what she had to say. She was exercised that I had impugned the biological worth of the microbial world. And she was literally standing up to defend that tiny unseen world, whose effects, though unheralded, are indeed manifest around us. I don't remember my rejoinders—if rejoinders there were: though I never shy away from an intellectual argument, there weren't many openings left in Lynn's verbal onslaught.

I've thought about that event many times through the passing years. Of course she was right: Dobzhansky must have had in mind the more conventional views on the "isolating mechanisms" that lead to reproductive isolation in natural populations. Why not admit that the effect of reproductive isolation apparently caused by a microbial infection in a lab was an interesting biological phenomenon in and of itself? A real effect of interspecific interactions of an up-close and personal kind? That it was, in fact, the very sort of intimate relation that had already long intrigued Lynn and led to her momentous conclusions about signal events in evolutionary history?

I hope I had the presence of mind to say something like this to Lynn, as she started to calm down that evening long ago at Amherst. For not only had I found myself in the presence of this already well-known and respected biologist, but I had learned something from her in that very first "interchange" (more like a harangue!).

As I grew to know her from that point on, I soon learned that the passion she evinced that evening was no unique event. There was no personal animus (of the sort so common in sharp exchanges in the academic arena); nor did there ever seem to me to be any hint of personal self-glorification as she continued through all those years to develop her ideas and speak forcefully in their behalf.

None of that: instead, the Lynn I met that evening, the true Lynn, was the epitome of intellectual honesty. She was single-mindedly and unselfconsciously propounding, defending, and promoting her view of the way things are in the biological world—most especially, of course, the incredibly diverse array of microbial organisms. Passionately. It's the ideas that really matter—and Lynn certainly had hers; they were novel and profound; and she simply wanted all the rest of the world to adjust their thinking to accommodate and embrace what she saw were the simple, beautiful truths that she had uncovered. Not a bad intellectual agenda! And, at least for the greatest, most important part of her life's work, she won!

As I grew to know her better, and become friends, I got to see her more personal side as well. We would talk on the phone (though her recalcitrance vis-à-vis email became something of an impediment to full communication in later years). We saw each other at meetings, of course. And once I had the pleasure of staying at her house in Amherst as I attended a meeting. That experience in itself was a revelation: for the house was full of family and I think other guests—there were a lot of people around, a phenomenon I have only really seen the likes of with Margaret Mead and Wynton Marsalis. She was truly magnetic.

But the time I remember best was at a meeting in Colorado, at least ten years ago. We agreed to meet for lunch. When I showed up, Lynn had a grandson in tow—and announced that she had promised him lunch at Chuck E. Cheese's. Would I mind?

Of course not. We sat in impossibly confining plastic chairs while her grandson played for awhile in a room full of plastic balls. We talked about many things—mostly science, probably some politics. It was fun. It was relaxing.

And that is why, when I heard of her death, that I wrote to her son Dorion Sagan that "everyone will say what a magnificent scientist she was . . . and they are of course right—she was an extremely dynamic and creative thinker. But she was also a lot of fun, which in my book is just as important!"

She was wonderfully passionate in all sides of her life!

Niles Eldredge, a paleontologist best known for proposing the theory of punctuated equilibrium with Stephen Jay Gould, has published more than 160 scientific articles, books, and reviews, including Reinventing Darwin *and* Dominion, *a consideration of the ecological and evolutionary past, present, and future of human beings.*

Lynn Margulis and Stephen Jay Gould

MICHAEL F. DOLAN

Salem State University in Massachusetts hosts an annual Darwin Festival during Darwin's birthday week in February. Students are excused from classes to attend lectures, workshops, and films. About ten years ago, Lynn and Stephen Jay Gould were keynote speakers on the same day. In his talk, Gould noted that Darwin and Abraham Lincoln were both born on February 12, 1809, and used this coincidence to recall the Battle of Little Round Top in Gettysburg as a metaphor for his views on the contingency of evolution. If the Mainers under Joshua Chamberlain hadn't dared their famous bayonet charge to hold off the Alabamans, and the Confederacy had captured the hill, the whole Battle of Gettysburg could have resulted differently, with subsequent changes to US history. Evolution also occurs in this contingent fashion, Gould argued, rather than in the iconic progression from apes to white Europeans and their relatives.

Prior to speaking Gould was asked to announce that Lynn Margulis would give a workshop called "What happens to trash and garbage?"

"I'm glad my good friend Lynn is handling that," Gould said with a smile, "because there are just some things I don't want to know."

Of course he missed the point. The workshop was not for sanitary engineers but for students to learn how microbes are involved in the global carbon cycle. Despite being one of our leading evolutionary biologists, with some appreciation of microorganisms, Gould had no ear for a discussion on the role of microbes, particularly microbial symbionts, in evolution.

Margulis and Gould were good friends and shared many attributes—their appreciation for the history of biology, their love of public speaking, their interest in communicating to the masses, their public presence. (Both were also a bit full of themselves. At the Salem event, when some students left the hall while he was speaking, Gould said, "Hey, where are you going? You can't hear me speak here every day.") They also both operated a bit outside the mainstream of modern science, funding their research by the honoraria they received for their lectures rather than applying for grants.

But they differed greatly in their perspectives. Gould was a paleontologist and had the fossil record as his database; however, each specimen he examined provided a limited amount of information. Lynn dealt with living organisms that could be harvested, cultured, and analyzed to the remotest molecule; however, she had to reconstruct the past using mostly contemporary life. Gould valued statistics; Lynn had contempt for statistics. Lynn centered her career on the study of symbiosis. Gould apparently was oblivious to it, or perhaps felt that it was only important during the Precambrian. He does not even include the word in the index to his monumental *The Structure of Evolutionary Theory*, published shortly before his death in 2002.

Gould had some appreciation for the importance of microorganisms. In his preface to Margulis and Schwarz's *Five Kingdoms* (1982), Gould wrote, "[The authors] have generated here the rarest of intellectual treasures—something truly original and useful. . . . It is remarkable that no one had previously thought of producing such a comprehensive, obvious, and valuable document."[1] While acknowledging being "disgracefully zoocentric," Gould praised the book's placing animals "into proper perspective on the tree of life." He would go on to accept that we live in the "age of bacteria," with a planetary system whose elemental cycles are run by bacteria, and in which plants and animals are recent, unnecessary arrivals. Gould devotes two pages (out of 1,300) to microorganisms in *The Structure of Evolutionary Theory*.

In that book he sided with molecular biologist Carl Woese rather than the more traditional biologist Ernst Mayr, writing, "Life's tree is, effectively, a bacterial bush. Two of the three domains belong to prokaryotes alone, while the three kingdoms of multicellular eukaryotes (plants, animals, and fungi) appear as three twigs at the terminus of the third domain." However, Gould tepidly endorses the usual "rather anthropomorphic perspective" of complexity, writing, "The sequence of bacterium, jellyfish, trilobite, eurypterid, fish, dinosaur, mammoth, and human does, I suppose, express 'the temporal history of the most complex creature.'"

This reveals the difference in the two scientists' perspectives. From a symbiogenetic approach, humans are the descendants of the product of a fusion of at least two organisms—a protomitochondrial host archaebacterium and the symbiotic invader that became the mitochondrion. But is that more complex than, for example, the ciliated protozoan *Myrionecta rubra* that contains, in addition to the two genomes just mentioned, a flagellate endobiont (with its own nucleus and mitochondrion), and which in turn harbors a red algal endobiont (with its nucleus, mitochondrion, and plastid)?

Perhaps nothing demonstrated the gulf between their two views of evolution more than the events at the eighty-fourth Nobel symposium in 1992, during which Gould became exasperated. Margulis gave a paper with Joel E. Cohen, "Combinatorial Generation of Taxonomic Diversity: Implication of Symbiogenesis for the Proterozoic Fossil Record," in which they argued that conventional species names, such as *Hydra viridis*, failed to recognize that the "organism" was actually two organisms—the animal *Hydra* and the green alga *Chlorella*—and that the whole organism (*Hydra viridis*) contained a combination of at least five genomes: the animal nuclear and mitochondrial genomes, and the alga's nuclear, mitochondrial, and plastid genomes. 2 + 3 = 5. They then made the speculative leap that led to Gould's exasperation.

Provocatively, and breezily, as was Lynn's style, she and Cohen speculated that one could derive all the diversity of beetles (Coleoptera) through a series of symbioses that led to speciation of the insects, or as Margulis and Cohen put it, "We predict that between 20 and 22 physiologically distinctive microorganisms (primarily bacteria and fungi) are regularly associated with coleopterans. Genomic combinatorics may explain why, as J. B. S. Haldane observed, God has expressed such an inordinate fondness for His most flamboyant morphotypes: His millions of species of beetles."[2]

Lynn later confessed to me that Gould was exasperated. "What is this, the Kabbalah?" he'd said. "What is this, numerology?" The gaping canyon between their two perspectives could not have been more apparent. Gould's ignored symbiosis, focused on statistics and the fossil record. Lynn's, extrapolating from extant symbioses, imagined a world in which symbiosis was the dominant engine of evolution.

Michael F. Dolan is an adjunct professor of geosciences at the University of Massachusetts, Amherst, and runs the Marine Biological Laboratory's NASA Planetary Biology Internship Program. As a student and colleague, he worked in Lynn Margulis's lab for nearly twenty years.

Too Fantastic for Polite Society:
A Brief History of Symbiosis Theory

JAN SAPP

Lynn Margulis's name is as synonymous with symbiosis as Charles Darwin's is with evolution. She developed symbiotic theory with regard to the origin of eukaryotic cells and later championed symbiosis as a central principle of evolution. In the 1970s, she articulated serial endosymbiosis theory (SET) according to which eukaryotic cells emerged from the multistage establishment of symbiosis. SET focused on three organelles: mitochondria, chloroplasts, and centrioles/kinetosomes, positing that these organelles originated as free-living bacteria that entered cells, became symbiotic, and never left. But just as there were evolutionists who preceded Darwin, there were symbiosis theorists before Margulis.

Research on the evolutionary importance of symbiosis has a long history that paralleled Darwinian evolutionary biology and the germ theory of disease. Though it emerged in the nineteenth century, it remained on the margins of biology throughout most of the twentieth century. Several scientists in various countries made proposals of eukaryotic organelle origins by symbiosis, sporadically and independently. German botanist Andreas Schimper proposed that chloroplasts might have originated as symbionts when he coined the word "chloroplasts" in 1883. The idea gained merit in the 1880s with evidence of "animal chlorophyll" in translucent animals such as hydra and sea anemones; such animal chlorophyll had been thought to have been generated by the animals themselves but was shown to be synthesized by internal symbiotic algae.

In Russia in the 1890s, Andrei Famintsyn conducted experiments to extract from animals symbiotic green and brown algae; he hoped to learn how to culture chloroplasts, grow them on their own. In 1905 Ernst Haeckel also postulated the idea that chloroplasts might be cyanobacteria. In the first two decades of the twentieth century, that idea was championed by Russian botanist Konstantin Merezhkovsky, who, in 1909, coined the word "symbiogenesis," which he defined as "the origin of organisms by the combination or by the association of two or several beings which enter into symbiosis."[1] Merezhkovsky explicitly denied that mitochondria originated as symbionts, but like others before him he proposed that nucleus and cytoplasm originated as a symbiosis of two different kinds of microbes.[2] Japanese zoologist Shozaburo Watase proposed that the nucleus and cytoplasm were symbionts, and that centrioles might be as well, in his Woods Hole Biological Lecture of 1893.[3] In his 1924 book *Symbiogenesis: A New Principle of Evolution*, Russian biologist Boris Kozo-Polyansky proposed that the motility of eukaryotes with cilia and flagella originated symbiotically.[4]

That mitochondria were symbionts that entered animal cells eons ago and played fundamental roles in tissue differentiation of their hosts was developed by French biologist Paul Portier, at the Institut Océanographique de Monaco, in his 1918 book, *Les symbiotes*. In *Symbionticism and the Origin of Species*, American biologist Ivan Wallin wrote of mitochondria as a heterogeneous population of symbionts repeatedly acquired. Like others of his generation, Wallin viewed chloroplasts as having been derived from differentiated mitochondria. Accordingly, the mitochondrial population in the egg represented a great number of bacterial strains repeatedly acquired in the course of evolution. Mitochondrial evolution and differentiation resulted in many other cell organelles, including Golgi bodies, cilia, and chloroplasts.[5]

In his 1913 *Problems of Genetics*, the famed British geneticist William Bateson, known for his non-Darwinian saltationist evolutionary views, proposed that new genes might be acquired exogenously from infections.[6] Similarly, Wallin proposed that repeatedly acquired mitochondria were the source of new genes.[7] In his view, evolution was governed by three principles: symbiosis was concerned with the origin of species, natural selection with their survival and extinction, and an unknown principle was responsible for the direction of evolution to ever more complex ends.[8]

Symbiosis applied to bacteria and their viruses, too, as both Wallin and Portier saw it. In 1917 French Canadian Félix d'Herelle reported on "an invisible microbe" that he named "bacteriophage" (bacteria eater) that decimated a colony of the dysentery bacillus. Two years later, he noticed that not all bacteria were destroyed by bacteriophages. Mixed cultures of phage and bacteria could be subcultured indefinitely, and there were transformations in the morphology and physiological properties of the infected bacteria. D'Herelle saw the analogy with lichen symbiosis. He referred to the mixed cultures as "microlichens." He also saw symbiosis as the main source of evolutionary change. Indeed, in 1926 d'Herelle declared that "symbiosis is in large measure responsible for evolution."[9]

Despite such claims for its fundamental place in life, microbial symbiosis was generally considered by biologists to be a rare, exceptional phenomenon. Although hereditary symbiosis was well documented, especially in insects, it was disregarded as being of little significance for genetics and evolution. The evidence of virus-harboring bacteria was rejected as unbelievable for decades before bacterial geneticists revitalized it in the 1950s; ideas that cytoplasmic organelles were symbionts were rejected out of hand until the 1960s and 1970s.

No Place to Sit

No matter how it was conceived, the integrative force of symbiosis in evolution remained close to the margins of polite biological society. There were several reasons for this.

1. The notion that bacteria could play any beneficial role in the tissues of their hosts was overshadowed by the success of the germ theory of disease, and it contradicted germ theory's tenet that asepsis (a germ-free state) is a characteristic of healthy tissue. Rather than viewing microbes from "the window of medicine," Portier said, he looked at "microbiology from the window of comparative physiology" and

envisaged "a new form of bacteriology: physiological and symbiotic bacteriology."[10] Similarly, Wallin commented in 1927, "It is a rather startling proposal that bacteria, the organisms which are popularly associated with disease, may represent the fundamental causative factor in the origin of species. Evidence of the constructive activities of bacteria has been at hand for many years, but popular conceptions of bacteria have been colored chiefly by their destructive activities as represented in disease."[11]

2. Hereditary symbiosis conflicted with nucleocentric conceptions of the cell and of concepts of the organism in terms of a single germplasm. In the opinion of those who developed the Mendelian chromosome theory, all cytoplasmic structures emerged during development under the control of genes in the nucleus. The leader of the *Drosophila* school of Mendelian genetics, T. H. Morgan, put this perspective in a nutshell in 1926, "In a word, the cytoplasm may be ignored genetically."[12] Classical geneticists defined heredity in terms of the sexual transmission of genes between individuals of a species. Infectious heredity or hereditary symbiosis was excluded.
3. Symbiosis conflicted with the basic tenets of the evolutionary synthesis—the neo-Darwinism of the 1930s and '40s—according to which natural selection acted on gradual transformations resulting from gene mutation and recombination within populations. Neo-Darwinian evolutionists' emphasis on an incessant struggle for existence within and between species implied that the establishment of stable hereditary symbiosis would be a rare exceptional occurrence. Known cases of symbiosis were relatively scarce, and they were treated as curiosities, "special aspects of life," of little significance for general biology.
4. Finally, one has to consider that symbiotic theories of cell organelles were highly speculative. They were beyond the range of experimental inquiry. One of America's premier cell biologists, E. B. Wilson, put it this way in 1925 when, considering the various proposals that cell organelles had arisen

> by symbiosis, he pronounced: "To many no doubt, such speculations may appear too fantastic for mention in polite society; nevertheless it is within the range of possibility that they may someday call for some serious consideration."[13]

Classical evolutionists' attitude about symbiosis was much the same as their attitude about the microbial world. They were generally just not interested. Margulis aimed to change all that. Darwin and his followers said virtually nothing about microbes and the origin of cells. Theirs was a two-kingdom world of plants and animals. Their evolutionary perspective was about the last 550 million years of evolution, thus essentially ignoring 85 percent of organismal evolution on earth. Their question was the origin of species, not the origin of kingdoms. Margulis's intention was to help bring the microbial world and therefore deep evolution to the fore when she championed the idea of five kingdoms proposed by plant ecologist Robert Whittaker. Subsequent disputes with the rise of microbial phylogenetics were not about fewer "kingdoms" but whether there were several more, as well as even higher taxa: "domains," as proposed in 1990 by Carl Woese and his colleagues.

While classical evolutionists were generally not interested in the evolution of cells, those who studied cells were generally not interested in evolution. Cell biology, like much of molecular biology, was highly removed from the study of evolution. This was the double bind that Margulis found herself in when she first theorized on the evolution of the eukaryotic cell. Cell biology courses of the 1970s did not typically include evolution any more than did molecular biology courses.

At the Crossroads

The turning point for symbiosis in cell evolution occurred in the early 1960s, when DNA and ribosomes were discovered in mitochondria and chloroplasts and with electron microscopy of the organelles.

Margulis first became interested in symbiosis and organelle origins when she was a master's student in zoology and genetics at the University of Wisconsin. She took Hans Ris's advanced cytology class. Ris reported on the evidence that "strongly supported the old idea of endosymbiotic origin of mitochondria and chloroplasts," at the International Congress of Cell Biology in 1960.[14] As Margulis recalled, Ris read aloud to the class from the third edition of E. B. Wilson's *The Cell in Development and Heredity* of 1925:

> Thus when Ris read to us: "More recently Wallin (1922) has maintained that chondriosomes [old name for mitochondria] may be regarded as symbiotic bacteria whose associations with other cytoplasmic components may have arisen in the earliest stages of evolution [. . . .] To many, no doubt, such consideration may appear too fantastic to mention in polite scientific society; nevertheless, it is in the range of possibility that they may some day call for more serious consideration . . . ", the course of my professional life was set forever![15]

Virtually all of those who showed, in the 1960s, that mitochondria and chloroplasts possessed their own DNA suggested that perhaps they were symbionts. When reporting electron microscopic evidence of chloroplast DNA in 1962, Ris and Walter Plaut at the University of Wisconsin commented: "The evolution of the complex cell, with its array of more or less autonomous organelles, from the simpler organization found in Monera [bacteria] is a question that has been neglected. With the demonstration of ultrastructural similarity of a cell organelle and free living organisms, endosymbiosis must again be considered seriously as a possible evolutionary step in the origin of complex cell systems."[16]

The next year when Sylvan and Margit Nass at the University of Stockholm reported evidence of DNA in mitochondria, they commented that a "great deal of modern biochemical and ultrastructural evidence . . . may be interpreted to suggest a phylogenetic relationship between blue-green algae and chloroplasts and bacteria and mitochondria."[17] The structural and functional similarities between mitochondria and chloroplasts and bacteria and blue-green algae were discussed and debated in more than fifty papers during the 1960s and 1970s. The International Society for Cell Biology hosted a symposium on this topic in 1966.

Margulis not only championed the idea of organellar symbiosis but gave it evolutionary meaning. Transforming idea into theory, she discussed mitochondrial and chloroplast origins together in a geological context, and in a way that had not been done before. After completing her master's degree in 1960, Margulis moved to Berkeley for her PhD, which she completed in 1965, on studies of chloroplasts in the algae of the genus *Euglena*. She subsequently studied microbial classification, symbiosis, and cytoplasmic organelles as a postdoctoral student at Brandeis University.

She also aimed to extend the symbiosis idea from mitochondria and chloroplasts to account for centriole/kinetosomes, as there were claims then that traces of DNA could be found in them, too. Electron microscopes confirmed the claim made decades earlier that centrioles were the same organelles as kinetosomes (or basal bodies) at the base of flagella (which they are still commonly called) and cilia, which she insisted on calling "undulipodia" to distinguish them from bacteria flagella, on the basis of their great difference in size and protein composition. Both comprise identical protein bundles; no comparable structures exist in bacteria. When the cilia or flagella project outward, they provide mobility, feeding mechanisms, retinal rods, and other structures that develop from them. The inner ends give rise to the mitotic filaments on which the nuclear material divides at cell fission, in synchrony with the division of the eukaryotic flagellum. Eukaryotic flagella and cilia were important inventions that conditioned eukaryotic evolution.

The undulipodia (cilia and flagella) of eukaryotes were very different from the flagella of bacteria. No homology could be found; they were not the same organs. But there was good evidence that the microtubule organelles centrioles and kinetosomes were homologous. To support her case that centrioles (and their sister structures, kinetosomes) arose as symbionts, Margulis pointed to their physical and behavioral similarities with spirochetes, which resemble cilia in their motion and can sometimes be seen attached to eukaryotic cells. Indeed they had been mistaken for cilia, when they were found attached to the protist *Mixotricha paradoxa* living in the hindgut of termites.[18] The morphological similarities between the locomotive organs of ciliated cells and spirochetes were indeed striking, and Margulis suggested that perhaps eukaryotic cells emerged when a primitive microbe had ingested a spirochete-like organism.

Under her first married name of Sagan, Margulis published her paper "On the Origin of Mitosing Cells" in 1967. It was by far the most daring and serious effort in pursuit of the symbiont hypothesis of that time. She explored the geochemical and paleontological record as well as the data of cytology, microbiology, and biochemistry upon which to situate a historical reconstruction of the eukaryotic cell by a series of symbioses. In the scenario she proposed, eukaryotes arose symbiotically from bacteria, which would have arisen between 4.5 and 2.7 billion years ago, and eukaryotic cells arose between 0.5 and 1.0 billion years ago, perhaps a billion years after the release of oxygen produced by cyanobacteria (blue-green algae). In her scheme of 1967, mitochondria came first. Aerobic bacteria-protomitochondria invaded or were then engulfed by a predatory microbe. Those ancestors of mitochondria proved to benefit their hosts because they made use of the increased levels of oxygen in the primitive atmosphere brought on by the advent of photosynthetic cyanobacteria.

The next step toward eukaryotes occurred when some of those mitochondria-containing microbes subsequently ingested motile spirochete-like organisms. The spirochete genes were eventually utilized to form the chromosomal centromeres and centrioles. The process that gave rise to mitotic cells, Margulis argued, had occurred many times in various primitive ameboflagellates.[19] During the course of the evolution of mitosis, which took millions of years, various lineages of protists were infected with cyanobacteria to generate the ancestral eukaryotic algae.[20] Later, in the early 1990s, Margulis switched this scenario to argue that motility through the acquisition of spirochetes came first, then the acquisition of mitochondria, and then plastids.

In the 1960s and '70s, a strong case for the symbiotic origin of mitochondria and chloroplasts had been built on the following facts: (1) that different degrees of integration of a symbiont into the life of a cell were known to occur; (2) that the plastids and mitochondria are separated by a double membrane from the rest of the cell; (3) that they reproduced by fission (like bacteria); (4) that their ribosomes were similar in size to prokaryotic ribosomes; and (5) that the organization of their DNA fibers as revealed by electron microscopy resembles the DNA organization in a bacterial genophore rather than the chromosomes of the nucleus of eukaryotes.[21]

Strong arguments could also be made for the alternate view that all cell organelles evolved autogenously. Geneticists had shown that cytoplasmic organelles such as mitochondria were not the whole organisms that some had previously assumed them to be. They were well integrated into the cellular genetic system, and only a small fraction of the genes needed for mitochondrial and chloroplast functions were actually located in the organelles themselves; most of the organellar proteins were encoded in the DNA of the nucleus. If organelle genomes had evolved at a slower rate than nuclear genomes, it was reasoned, one could explain the similarity of organelles and prokaryotes without recourse to symbiosis. The features common to bacteria and blue-green algae on the one hand and mitochondria and chloroplasts on the other would then be due to "retained primitive states."[22]

Proposing a non-Darwinian evolutionary process only made matters worse for the exogenous argument. Rapid evolution by symbiotic leaps and not gradually through gene mutations and selection was tantamount to creationist thinking, as Thomas Uzell and Christina Spolsky saw it:

> The endosymbiosis hypothesis is retrogressive in the sense that it avoids the difficult thought necessary to understand how mitochondria and chloroplasts have evolved as a result of small evolutionary steps. Darwin's *On the Origin of Species* first provided a convincing evolutionary viewpoint to contrast with the special-creation position. The general principle that organs of great perfection, such as an eye, can evolve provided that each small intermediate step benefits the organisms in which it occurs seems appropriate for the origin of cell organelles as well.[23]

Non-Darwinian was beside the point; the only real question was how to test it. There was no definitive proof for either side. Famed microbiologist Roger Stanier quipped in 1971, "Evolutionary speculation constitutes a kind of metascience, which has the same fascination for some biologists that metaphysical speculation possessed for some medieval scholastics. It can be considered a relatively harmless habit, like eating peanuts, unless it assumes the form of an obsession; then it becomes a vice."[24]

Margulis told me she thought that Stanier was referring directly to her. Perhaps he was. His views were certainly shared by most cell biologists and molecular biologists, too: one could not know the origin of cells, any more than one could know the origins of the genetic code. These were metascientific questions beyond empirical science. Margulis agreed that "historical theories, which necessarily treat complex irreversible events, can never be directly tested." She argued that evolutionary biologists were in the same logical predicament as historians, and "can only present arguments based on the assumption that of all the plausible historical sequences one is more likely to be a correct description of the past events than another."[25]

New Methods

All of this changed when new concepts and methods of molecular phylogenetics emerged, which closed the basic argument over the origin of mitochondria and chloroplasts. At the University of Illinois, Carl Woese developed methods based on comparisons of ribosomal RNA sequences to reveal bacterial evolutionary relations, hitherto deemed impossible. Woese's methods led to his proposal of three domains: the Archaea, the Bacteria, and the Eukarya. Margulis and Woese were never to see eye to eye about a taxonomy based on three domains above the level of kingdoms. But the methods Woese developed were crucial for demonstrating the symbiotic origin of organelles. Comparing ribosomal RNAs of chloroplast, mitochondrial, and nuclear origin with one another and with different kinds of bacteria closed the main controversy about their origin.[26] Based on rRNA sequence comparisons, mitochondria and chloroplasts were of alphaproteobacterial and cyanobacterial origin respectively.[27]

Margulis brought the new molecular evidence together, drawing on the many facets of the evolutionary drama, in her *Symbiosis in Cell Evolution* of 1981. That book, today a "citation classic," heralded what Yale University biologist and fellow National Medal of Science winner George Evelyn Hutchinson, himself considered a father of ecology, called "a quiet revolution" in biological thought. But the revolution for Margulis was not complete.

First, the spirochete origin of the genes for motility and cell division remained in doubt. Few biologists supported it; Margulis was the only vigorous promoter during the 1980s and 1990s. Centrioles did not divide as did mitochondria and chloroplasts, and existence of DNA within basal bodies and centrioles was questionable. The evidence for DNA in centrioles had been on-again, off-again since the 1960s, but it was effectively refuted in the 1990s by evidence from electron microscopy and molecular hybridization, which indicated that genes affecting centriolar/kinetosome function are located in the nucleus.[28] Still, the lack of DNA in those organelles was not proof of their autogenous origin.

It was still possible that spirochete genes had been transferred to the nucleus. By the end of the twentieth century, genomic comparisons of ancient genes indicated that the nucleus of all eukaryotic cells were of at least three lineages: (1) informational genes, those involved in transcription and translation that seemed to be of archael origin, (2) metabolic genes that seemed to be transferred there from mitochondrial genomes (and chloroplasts in plant and algal cells) as expected, and (3) other genes found in the nucleus that were not obviously transferred from mitochondria. The latter genes encoded proteins that give eukaryotes their distinctive cellular character: actin, tubulin, and the main components of the microtubules of their cytoskeleton.[29] Could these be the long-sought spirochete genes? While many cell origins researchers agree that these genes may well be vestiges of another symbiosis, they insist that these genes are not of spirochete origin.[30] Margulis held fast to her spirochete theory.[31]

There was still another aspect to the symbiogenesis revolution that Margulis envisaged: symbiosis is the primary mode of evolutionary change in all organisms.

Bellagio, Italy, 1989

In 1989 I received a letter from Margulis inviting me to a weeklong workshop called "Symbiosis as a Source of Evolutionary Innovation" at the Rockefeller Foundation Bellagio Study and Conference Center, Lake Como, Italy. That was, I believe, the first meeting of its kind ever held. About twenty isolated researchers from North America and Europe working independently on diverse phenomena of symbiosis converged on Lake Como that June. Few symbiosis researchers knew one another then. There was no symbiosis society as there is today. (Margulis later cofounded the International Symbiosis Society.)

Margulis had organized the Bellagio meeting around the phenomena of symbiosis and their evolutionary significance. The examples were plentiful and diverse. Some symbionts were inherited vertically through the cytoplasm of the protist or through the eggs of insects.

Other examples were acquired horizontally from the environment. They include the Hawaiian squid that capture luminescent bacteria from the sea and keep them in special light organs, as Margaret McFall-Ngai explained. Even our understanding of the bacterial world of evolution was changed at that meeting. Sorin Sonea introduced the bacterial world as a sort of superorganism, a worldwide web of rapid adaptive evolutionary changes resulting from horizontal gene transfer—a view that has since been confirmed by microbial phylogenetics.[32] It was a whole other world, a dazzling panopoly of natural history that Margulis brought together. Everyone's views of nature were transformed.

The word "symbiosis" itself became a subject of intense discussion at that meeting, too. Did "symbiosis" mean mutualism? That was certainly how many biologists then and now use the term. Margulis was rightly skeptical of cost-benefit analysis of symbiotic interactions, the just-so stories created, and their (anthropomorphic) typology as mutualism, parasitism, or commensalism. She also had no time for sociobiology and its game theory void of physiology and ecological context. She made it clear that she wanted "symbiosis" to signify contiguous relationships with physiological and morphological effects on microbe and/or host.

Although Margulis did not know it then, hers was actually the original meaning of "symbiosis" as defined by Heinrich Anton de Bary in 1878. De Bary used the term to describe "the living together of unlike named organisms" and he recommended its study as a mode of discontinuous evolution to complement gradual continuous evolution of the Darwinian kind. De Bary pointed to the dual nature of lichens and of fungi living in the tissue of plants that seemed to cause no harm. Mycorrhizae in the roots of trees, nitrogen-fixing bacteria in the root nodules of legumes, and microscopic algae living in the tissue of translucent animals such as sea anemones, coral, and even some flatworms were well known by the early twentieth century. So, too, was hereditary symbiosis in some insects. But they were all considered curiosities. It was difficult enough for humans to cooperate, it was said; many biologists found it impossible to believe that different species of lower life-forms could live together in a stable relationship. Microbes were parasites; they were there to steal the rightful inheritance of their host.

Suffice it to say, de Bary's vision was not shared by many biologists over the following hundred years or so. But it was exactly what Margulis independently envisioned, and the reason she had brought us to Bellagio twenty-three years ago. The idea that symbiotic events occurred but were anomalous persisted to the end of the twentieth century.

John Maynard Smith was a participant at the Bellagio meeting in 1989; he was as impressed with the diverse phenomena as any of us, but he was critical of evidence of any mode of evolution that was not gradual and driven by selection. Indeed, so was virtually everyone else. The classical architects of the so-called evolutionary synthesis of the mid-twentieth century had long agreed that gene mutation and recombination within species were the only sources of evolutionary innovation by natural selection. That the principles of microevolution (within species) were the same as those of macroevolution (above the species) was central to that synthesis. All other claimed modes of evolutionary change were erroneous, especially saltationism. There were no new mechanisms of evolution. "If it's new, it isn't true" was their silent mantra: evolutionary theory is effectively over; its mechanism has been discovered; everything can be understood by extant principles of population genetics.

One day, after listening to several speakers on the effects of microbial symbionts on animals, Maynard Smith commented that the phenomena discussed were not about macroevolution unless one could show that the behavior of the organism is changed. The next speaker was Paul Nardon from Lyon, France, who talked about microbial symbionts in weevils, which when experimentally removed rendered weevils unable to fly! We all turned to Maynard Smith. Good enough? Nardon interpreted the phenomena of hereditary symbiosis as a form of neo-Lamarckism: the inheritance of acquired characteristics. I thought we were transcending both neo-Darwinism and neo-Lamarckism.

Zoocentric Biology

Ten years later, in 1999, Maynard Smith and Eörs Szathmáry asserted that "transmission of symbionts through the host egg is unusual." And when it did occur the microbes should be considered slaves.[33] Such views were based on theoretical assumptions about the evolution of cooperation, based on game theory and selfish genes. Despite what those assumptions led one to believe, hereditary symbiosis *is* prevalent, especially among insects. Bacteria of the genus *Wolbachia* are inherited through the eggs of at least 25–75 percent of all insect species and those of nematodes, too. And far from being slaves, they manipulate the development of their hosts, causing parthenogenesis and cytoplasmic incompatibility, and can turn functional males into functional females.

It would be a mistake to think that only neo-Darwinians such as Maynard Smith were skeptical of the pervasiveness of symbiosis as a source of evolutionary innovation. Stephen Jay Gould had also been invited to the meeting. Margulis read aloud his letter explaining that he declined because he was busy writing about macroevolution. Gould's book *Wonderful Life* appeared later that year. It was an excellent read, but his comments on microbes and about symbiosis had all the earmarks of a classical evolutionist. When writing against "evolutionary progress," he said that nothing much happened in evolution for billions of years. After all, he said, animals did not emerge until 550 million years ago or so. Then, turning to the origins of eukaryotes and the symbiotic origin of mitochondria and chloroplasts, he said he was "entering the quirky and incidental side" of evolution.[34]

Those comments typified what Margulis was up against—her struggle against what she called "zoocentric biology" that typified classical evolutionary theory, nonmicrobial biology. One cannot underestimate the lack of interest in the microbial world then, which persists today among classical Darwinian evolutionists, their one-genome, one-organism conception of the individual derived from the erroneous assumption that symbiosis as a source of evolutionary innovation is an exceptional phenomenon.

Jan Sapp, in the Biology Department at York University, is a world expert on the history of symbiosis theory.

Kingdoms and Domains:

At Work on the Linnaean Task

MICHAEL J. CHAPMAN

Lynn Margulis was proud that her house at 20 Triangle Street in Amherst, Massachusetts, was next door to the Emily Dickinson homestead. She loved to spend time at home cooking, entertaining, conversing, reciting Dickinson's poetry from memory, and, above all else, working in the dim, steeply gabled, third-floor attic she had converted into a furnished office space. Lynn's home office was a sanctuary, an effect intensified by the classical music she often had playing.

Whenever I came to work for Lynn in Amherst, I would stop first at 20 Triangle Street and spend an hour or two in her office, the two of us glued to some micrograph on a computer screen while she talked and I listened. In one-on-one conversation, Lynn had a way of making people feel special and indispensable. She would fix me with her steely gaze and say, "You know, Mike, you are the only person in the world I can talk to about this stuff!" And so, when she proposed in 2006 that the two of us collaborate on a fourth edition of her influential *Five Kingdoms* reference volume, how could I refuse?

Lynn and I shared an interest in taxonomy because of its connection to evolution. The Linnaean task, that of naming and classifying all life on earth, is an enormously complex and often counterintuitive set of challenges. Aristotle, for example, initially classified plants as either trees, shrubs, or herbs; we now know that some plant families, such as the roses or legumes, have arboreal, shrubby, and herbaceous representatives, and that half of flowering plants (about 125,000 species) are of hybrid origin. Bacteria routinely exchange genes between what are called different species, an act that stretches the classical definition of a species as a group whose members can mate successfully only with one another. Many animals, such as Portuguese man-o'-war (*Physalia*), are now known to be not single individuals but colonial associations of specialized zooids, some of which trap prey, some of which produce sperm or eggs, one of which floats, and so forth. The individual zooids cooperate rather than compete with one another, and selection acts on the whole colony at once. In short, there are many other modes of evolution than classical Darwinism—gradual accumulation of mutations coupled with natural selection—can account for.

In the fourth edition of *Five Kingdoms*, which came to be called *Kingdoms and Domains*, Lynn and I wanted to catalog the most recent findings and taxonomic arrangements of the phyla of life on earth. We also wanted to draw attention to combinatorial modes of evolutionary change, such as those that gave rise to *Physalia*, transformed bacteria, and allowed plants to hybridize. Lynn's greatest criticism of the predominant form of taxonomic data summary—what is known as the phylogenetic tree—is that a tree is an inadequate representation because its form is only outward-branching and never reticulate, or net-shaped. In other words, the phylogenetic tree cannot record speciation events via symbiogenesis, hybridization, or lateral gene transfer.

The most comprehensive phylogenetic tree of life was formulated in 1990 by Carl Woese,[1] who pioneered the technique of 16S rRNA (ribosomal RNA, the main structural component of ribosomes) sequence comparisons to draw conclusions about relatedness between groups of organisms. Because mutations accumulate in the rDNA (ribosomal DNA, the genes which encode rRNA) at a constant rate[2] and because ribosomes, as protein factories being so essential to the life of the cell, are unlikely to successfully undergo lateral gene transfer, 16S rRNA comparisons can be used to construct a phylogenetic tree.

Woese's tree divides life into three great domains: Bacteria, Archaea, and Eukarya. Note that Woese divides the prokaryotes into two domains; Archaea are designated as prokaryotes that are not bacteria. This division of the prokaryotes did not sit well with many prominent scientists, including Ernst Mayr, Salvador Luria, and Lynn herself. Lynn objected that no matter what the 16S rRNA sequences might indicate, the Archaea still had a great many more cellular features in common with the Bacteria than either group did with eukaryotes. She maintained that in order to represent evolutionary relationships accurately, we need to consider everything we know about the species: cell biology, symbiotic composition, and life cycle, for example, in addition to its 16S rRNA sequences.

Despite the mainstream scientific community's acceptance of Woese's three-domain scheme, Lynn and I did not adopt it for *Kingdoms and Domains*. We maintained the organization of the third edition in our fourth-edition reference volume: five kingdoms within two superkingdoms.[3] Superkingdom Prokarya includes single-celled or multicellular organisms lacking nuclei, microtubules, or membrane-bounded organelles and is further divided into subkingdoms Archaea and Eubacteria, both of which existed, according to the fossil record, from 3,900 million years ago to the present. Superkingdom Eukarya includes organisms that arose via symbiogenesis and belong to the kingdoms Plantae, Animalia, Fungi, and Protoctista.

Lynn and I felt that by taking into account all selectively important aspects of an organism's biology as well as its molecular sequence data, we were giving a more informative, more accessible account of nature than can be gleaned from Woese's tree. Most important, our scheme incorporates data on symbiogenetic origins of the phyla, which a branching, nonreticulate tree cannot. To cite an analogy from information science, the early medieval libraries had their holdings arranged in order of acquisition, and that is just as valid an organizational scheme as the Dewey decimal system. But which is more accessible?

Admittedly, this philosophy brought us into conflict with mainstream taxonomy. For example, we group the chytrids (phylum Chytridiomycota; fungal ancestors) and green algae (phylum Chlorophyta; plant ancestors) into kingdom Protoctista, because chytrids, unlike fungi, possess an undulipodiated zoospore, and the algae, unlike land plants, do not develop from an embryo. The most recent sequence data have shown that chytrids are, in fact, fungi, and green algae are plants. Multigene analyses have shown that undulipodiated zoospores and enclosed embryos are developmental stages and should not be taken for taxonomically significant characters. Yet Lynn always preferred the classical findings of cell biology, anatomy, and physiology to molecular biology. So in most such instances of conflict as the chytrids and green algae, we would simply report the molecular findings in the text and keep the grouping as it had stood in the third edition.

Lynn had hundreds of scientific contacts worldwide, so our next steps were clear: to contact world experts on all the various phyla and then to incorporate new findings, figures, and so forth, to update the third edition of *Five Kingdoms*, which was now over ten years old. This difficult, tedious, but important job was undertaken with cheerful aplomb by our editorial assistant, Kendra Clark. As each rewritten chapter came in, it would cross my desk for review. One change I frequently made was to substitute the word "undulipodium" for "flagellum" in the chapters on eukaryotes. Lynn insisted on this substitution because bacteria also have a whiplike motility organelle called a flagellum, with a structure vastly different from the eukaryotic flagellum, or sperm tail. Other cell biologists derided Lynn's use of "undulipodium" since it is a cumbersome term compared to "flagellum," but Lynn insisted on the term "undulipodium" for eukaryotic flagellum or sperm tail for the same reason she advocated the five-kingdom, reticulate phylogenetic tree: in order to avoid confusing students.

I often think back on the two years during which *Kingdoms and Domains* was in its editorial phase—the long, pensive afternoons spent working on Lynn's computer in her spacious, book-lined office in Morrill Science Center at the University of Massachusetts, Amherst. Lynn loved books and had accumulated a formidable library of tertiary scientific literature, so that the answer to practically any question was there on the shelves or in the hundreds of offprints of Lynn's own papers stored in filing boxes beneath the tables. My formal training is as a plant biologist, but in the course of my work with Lynn, my interests broadened to include all of life, especially Lynn's favorite organisms, the protists. These "water neithers"—neither plant nor animal nor fungus yet still eukaryotes in their own right—include some of the most beautiful and fascinating life-forms on earth, such as the termite hindgut organisms, which are poisoned by oxygen and make their living by helping the termite digest wood, or the highly motile green ciliates *Paramecium bursaria* and *Stentor coeruleus*, which incorporate living, photosynthetic algae (*Chlorella*) in their bodies and survive by photosynthesis. During this time I became interested in how eukaryotes evolved from prokaryotes and how likely it was that some approximation of the basal eukaryotic cell could be found in kingdom Protoctista.

Gradually, all the revisions from the various contributors found their way back to my desk, the necessary figures were assembled, and the raw materials for *Kingdoms and Domains* took shape. Our editors at Elsevier Academic Press were becoming more and more insistent that the book move into production. It was at times difficult to coordinate with Elsevier, since our acquisitions editor was in Oxford, our development editor in San Diego, and our production editor in Chennai, India. All of these positions seemed to have high turnover rates, and we had endless difficulties in our attempts to contact the unknown person whom Lynn called "the Boss." Against all odds, we finished preparing the volume just in time for Elsevier's deadline, and on the night of December 19, 2007, with Kendra scheduled to deliver her baby the following day, another superb editorial assistant, Idalia Rodriguez, and Lynn put together a sixty-five-pound box of primary manuscripts, original micrographs, diagrams, and drawings—the raw materials for the book—and shipped them off to Chennai.

Over the next year ensued a series of small frustrations and disappointments, as Elsevier rushed to production a first effort that was shot through with production errors: the appendix appeared in the front of the volume; diagrams were presented upside down; micrographs were transposed and mislabeled. At Lynn's insistence, the book was recalled from shelves and corrections made according to our specifications. What finally hit the shelves was much improved but still not perfect.

It is perhaps not surprising, considering Elsevier's far-flung operations and the high turnover rate among their employees, that the sixty-five pounds of primary scientific materials that Lynn and Idalia had sent them on that winter night in 2007 were ultimately lost. That box contained precious original scientific results that had either been produced by Lynn herself or donated by one of her illustrious friends. Without that box, it will be almost impossible to publish an updated fifth edition when the time comes. Lynn spent the last years of her life embroiled in a dispute with Elsevier over the box's whereabouts, demanding just compensation despite realizing, I think, that it was a quixotic venture.

The finished product, *Kingdoms and Domains*, is nevertheless a worthy contribution to the Linnaean task. The "two-superkingdoms" organizational scheme (Prokarya and Eukarya) offers a reasonable synthesis with Woese's three domains, since Prokarya remain subdivided into Archaea and Eubacteria, while still acknowledging the evolutionary importance of symbiogenesis (the major difference between Prokarya and Eukarya) that Woese ignores. Whenever I leaf through my own copy, I am transported back to those sunny afternoons in Lynn's gabled third-floor office sanctuary, listening to classical music, ideas whizzing through the air.

Michael J. Chapman is coauthor with Lynn Margulis of Kingdoms and Domains: An Illustrated Guide to the Phyla of Life on Earth, *with an introduction by E. O. Wilson.*

The Battle of Balliol

MARTIN BRASIER

On a chilly evening in May 2009, I was strolling down Broad Street in Oxford toward the Gothic portals of Balliol College, famous as the onetime haunt of British prime ministers. Its great oak-paneled doors were barely ajar, and a notice was propped against a forlorn-looking empty chair inside: "No admittance to the public. Private function within." Feeling less certain than before, though I was expected at the function, I crept through the gap and past the porter's lodge to sidle into this secret inner world of Balliol scholars. Crossing a dark outer courtyard, I slipped into the long green space of Balliol's inner quad and stepped toward room 23. A prime number. And a debating chamber.

A hubbub rose up the stairwell. This chamber proved to be a subterranean cavern, bubbling like a hydrothermal vent. And two great thinkers on evolution were about to feed upon this energy, in a habitat set below the reaches of sunlight. On one side stood Lynn Margulis, author of *Symbiotic Planet* and the greatest living proponent of symbiogenetic theory. On the other side stood Richard Dawkins, author of *The Selfish Gene* and the most public proponent of neo-Darwinian theory.

Room 23 was crammed with scholars eager to witness their champions in verbal combat. That is because Lynn Margulis and Richard Dawkins had learned different lessons from the same book of life. One of them saw networks. The other saw hierarchies.

I glanced around the room as the noise died down. Here could be found biologists seated on substrates, geologists camped uncomfortably above stone floors, and physiologists arrayed like neural networks outside the doors beyond. In the chair was Denis Noble, author of *The Music of Life* and a famous exponent of systems biology. Microbiologist Steve Bell and myself—a geobiologist—were also invited to join the platform. After just a few words, cameras began to roll.

Lynn Margulis stepped into the limelight and eagerly introduced her case. She laughingly teased the students for being spoon-fed a diet of greasy genes and dangerous DNA. Glowing with knowledge garnered from the living world, Lynn was soon telling us of her conviction that nearly all major speciation events that had ever taken place in evolution were the product of a kind of dangerous liaison—a symbiosis between two distantly related organisms that wantonly swapped their genetic information to form completely new genetic strains. Taking us next to the shores of Brittany, she presented the evidence of a little flatworm that grows its own garden of microscopic algal cells, captured within its tissues like a biodome. To illuminate her theme, a video began to play, shot mainly in black and white like a French film noir. We watched these movies in darkened room 23, while Lynn glowed with enthusiasm, exposing prime liaisons of the microscopic world.

We learned of spirochetes that gyrated—in this case to a kind of disco music—until they dropped. Of protozoans that flirted with these tiny spirochetes, often to catchy tunes. And of termites that had fallen hopelessly in love with whirling trypanosomes (a group of parasitic protists, one kind of which causes sleeping sickness in humans), all to the sound of strings. The whole living world seemed to be dancing, flirting with flagellates, or banging away with bacteria. Step by step, we were introduced to this closet world of protoctists thriving in slimy layers. Fed with characters from a biofilm noir, the biosphere seethed with networks unseen. And from such liaisons had arisen cell organelles—undulipodia, mitochondria, and chloroplasts—that together formed partnerships that were greater than the sum of their parts. Even our own cells were communes. Lynn, our speaker, who was then Eastman Professor at Balliol, went further by saying that nearly all forms of biological innovation could not have arisen by natural selection working on random mutations. More likely, they arose from networking. This Margulian thought is one of evolution's most beautiful ideas.

Richard Dawkins stepped forward into the fray. Silver-haired and clad in a light summer suit, he calmly explained that genes are potentially immortal. They are the only part of a cell that can live forever. As such, they are the ultimate carriers of biological information and therefore the ultimate units upon which Darwinian natural selection itself is obliged to work. Genes are regenerators par excellence. And they sit atop a hierarchy that runs from genes, via cells, through organelles, to individuals and thence to communities. The first Simonyi Professor for the Public Understanding of Science at the University of Oxford from 1995 to 2008, Richard went on to suggest that the assembled supporters of symbiogenesis didn't go far enough. With a twinkle in his eye, he suggested that the relationship between neighboring genes strung out along a chromosome could be regarded as yet another example of symbiosis. It was a fine oratorical ploy.

Among the last to speak was physiologist Denis Noble. Now, Denis has a remarkable capacity to captivate an audience—evolutionary or otherwise—by singing mournful songs in ancient Occitan to the tune of his Spanish guitar. (With Lynn, I was once privileged to be serenaded in this way.) Denis uses this troubadour's trick to remind us that both cells and organs are like the orchestral instruments and that the metabolisms of organisms going about their daily business compare with the sounds that we hear in our heads. Little of this music would survive down the centuries, of course, were it not for the written score sheet, which is like the genes and DNA. Transmission of music can also take place without the need for written music—from one musician to another.

Denis explained that an odd realization struck him while he was working as a cardiac physiologist: all cells in the human heart share the same genetic code, but they vary enormously in the shape and form, ostensibly to protect the heart from the transmission of unwanted rhythms. These cells seem to know where they are situated within the heart and invariably adopt the shapes required of them. So far as we know, no messages are encoded within the genes that could give out the instructions needed for the cells' shapes and positions. Instead, it seems, the messages are epigenetic, transferred by the cells themselves. This mode of transmission could be regarded as shockingly Lamarckian, but the paradox is hard to ignore. For Denis Noble, therefore, the ultimate unit of selection is the musical instrument: the cell.

The music of life is a beautiful analogy. But Richard Dawkins came back to argue that we need to differentiate here between the unit of replication—DNA or RNA—and the unit of reproduction: the cells of an organism. We were here plied with another colorful metaphor, this time a fire spreading across a dry grassy savannah. The main fire could send up sparks, and these might start little satellite fires, like reproductions of a parent. Here and there, their flames might be blue or yellow, owing to the local abundance of copper or sodium in the soil beneath. If the parent flame were intrinsically blue, it would need to have a replicator to ensure that the daughter fires were also blue in color. To Dawkins's eyes, DNA is like that replicator mechanism. Without it, there is no replication, only drift.

So what might the fossil record reveal for us here? This allowed me to argue my case as a geobiologist, that the fossil record is much better than Darwin believed. Indeed, it shows us things no biologist would dare to conjure —until the recent discovery of complex system behavior, that is—rapid changes in evolution such as the Cambrian explosion, characterized by the sudden profusion of new animal forms such as trilobites 530 million years ago, or mass extinctions in which global communities have vanished with hair-raising speed. While there seems little hope of tracing DNA down through the fossil record directly—the molecules degrade too quickly—we also have good evidence for the evolution of symbiogenesis in the forms of fossil foraminifera. These amoeba-like protozoans—called forams, for short—have evolved delightfully intricate shells, which fall upon the seafloor in abundance. Foram shells can be wonderfully diverse in shape, as well as in form—each species adapted to a limited range of seafloor conditions. Luckily for us, their shells can also contain the imprint of both the physics and chemistry of the waters in which they lived. It is with shells like these that we now model the history of climate and of reef evolution. They are the fruit flies of the fossil world.

What matters here, though, is that many living forams have also benefited from taking in red, green, or brown algal unicells as symbionts. Over tens of millions of years, foram hosts gradually changed the size, shape, and properties of their shells in ways that tended to maximize their efficiency for photosymbioses, their partnerships with photosynthetic algae. Their dead shells now contribute to *Globigerina* ooze of the deep seafloor, which is ultimately uplifted to make land. And they make up both reef rock and beach sand. Indeed, swaths of the planet are shaped by the remains of symbioses like these. They provide us with the geological context for symbiogenesis, to which I shall return below.

So who won this debate at Balliol, you may ask? The answer seemingly comes down to which of three units of selection—the gene (Richard Dawkins), the cell (Denis Noble), or the community of organisms that make up a living entity (Lynn Margulis)—has been more important for the history of life. And part of this hangs, of course, on what we mean by "important" and what we mean by the "history of life." Richard Dawkins agreed, for example, that symbiosis exists, but he went on to defend his position by saying that symbiogenesis is rare as an evolutionary process.

This is a highly debatable point because most organisms are microbes, most genetic transfer within the biosphere takes place among bacteria and archaea, and the vast majority of such microbes are found within symbiotic associations. Indeed, many are utterly symbioholic. Further, we can argue that bacteria, archaea, protoctists, and fungi hold a commanding influence over the planetary cycles of water, carbon gases, and nutrients. The debate then hinges around which of two life strategies has been the winner in a great planetary game: hierarchies, à la Dawkins; or networks, in the Margulian mode?

As a geologist concerned with larger-than-life and macrocosmic processes, I have struggled to gain a global perspective on such matters. My quest began in 1970, with a year spent as ship's scientist aboard HMS *Fawn*, mapping microbes and mangroves of the Caribbean. This was followed by several decades exploring the context for the Cambrian explosion around the world. And now I work on the earliest signs of life. From these travails, I would argue that each of the most important steps in evolution—the origins of cellular life (before circa 3,500 million years ago), of eukaryote cells (c. 1,800 Ma), of animal guts (c. 580 Ma), and of plant roots (c. 450 Ma)—has arguably involved revolutions in symbiogenetic networks on the grandest scale. RNA made new music with membranes. Purple bacteria sidled up to cyanobacteria. Animal cells entertained forms of intestinal bacteria. And plant cells fell into step with new forms of fungi. It was arguably such networking revolutions in the deep history of life that made our planet habitable.

This vision needs vastly more research, but it points toward fruitful questions, rich with ramifications.

Charles Darwin, in whose homage this meeting was held, was never really a reductionist, of course. Instead, he quizzed the biosphere through a distinctively concave lens, giving to life its proper geological and cosmic perspective. His pioneering work on coral reefs and then on earthworms in soils has provided lessons for us all. Darwin was, in spirit, a holist and a geobiologist, and Lynn Margulis (a Darwin-Wallace Medalist in 2009) was his distinguished successor.

When we now look at oceans, atmospheres, and soils in terms of their evolution in deep time; or we fret about our planet's fragile habitats; or we look for life beyond Earth—it is to microbes, their symbioses and their networks, that we increasingly turn, in humility, for instruction. And it was Lynn Margulis who helped us to turn that corner.

Martin Brasier is professor of paleobiology at the University of Oxford and author of Darwin's Lost World *and* Secret Chambers: The Hidden History of the Cell.

Science, Music, Philosophy:
Margulis at Oxford

DENIS NOBLE

"Are *you* Denis Noble?" An earnest enquiring face appeared across the elegant lunch table in the senior common room of Balliol College one day in the autumn of 2008. Eventually, I became accustomed to the unusual way in which Lynn would place emphases in her speech, but what came across immediately during this first interaction was that she was on a mission. The emphasis conveyed an urgency and necessity. She had already decided that we needed to meet.

I am ashamed to confess that I did not even know who she was! I knew, of course, about the idea that mitochondria and other organelles had originated from bacteria. I had also made my own forays into aspects of evolutionary biology while writing my book *The Music of Life*, a watershed experience during which I discovered that standard evolutionary theories make little sense in physiological science. If organisms were merely vehicles for genes that mutated entirely at random, then physiology would be largely irrelevant to the major processes in creating new species. That seemed to me to be strange. The whole point of an organism is its phenotype. Genes are simply part of what makes the phenotype possible. Like any "code," it must be read, and the reader must be at least as important as the data. In fact more so, since it, and it alone, can be said to introduce meaning to the data sequences.

I was therefore already coming to the conclusion that something must be incomplete or even fundamentally wrong with the modern synthesis (neo-Darwinism), but there were still large lacunae in my knowledge. During the year in which she was the Eastman Visiting Professor at Oxford University, some of those gaps became filled, and I now regard it as a great privilege that they should have been filled by someone of her distinction and achievements. I quickly learned how extensive symbiogenesis may have been in evolution, that the "tree of life" was better described as a network, and that she also didn't think much of neo-Darwinism.

I should have known who Lynn was, of course. Because she was a member of the National Academy of Sciences and a winner of the National Medal for Science, presented by President Clinton, there was little wonder she had been chosen as the Eastman Visiting Professor at Oxford. But my period as a member of the selection committee for the Eastman Chair had finished before she was chosen. The list of her predecessors is very distinguished, many of them Nobel Prize winners. That is the level of distinction that the Eastman Chair requires, though of course there are not enough such prizes for all who deserve them.

Perhaps my ignorance of her work was an advantage. I doubt whether she knew much of mine, except for *The Music of Life*. Each of us could explore our common interests from relatively naive positions. Something I learned quite quickly was that Lynn Margulis warmed to people's questions. Her eyes lit up when I confessed that I no longer knew what a gene was and that, whatever it was, I thought that selfishness was an inappropriate metaphor. To a physiologist, the striking thing about genes and their products, RNAs and proteins, is that they act in large cooperative networks to produce physiological functions. The networks reduce the influence of any individual genes on functionality by buffering the organism against such changes.

During those first encounters, Lynn may not have known that I had interacted with Richard Dawkins over many years, in fact even before *The Selfish Gene*. On the publication of that book, I organized a debate in the Balliol Graduate Centre at Holywell Manor in which he was asked to respond to two philosophers, Anthony Kenny and Charles Taylor. Tony Kenny threw out a challenge, saying that by knowing the letters of the alphabet one would not be entitled to say that one could understand the works of Shakespeare. As I recall it, Richard's reply was to the effect that he was just a scientist, interested only in truth. I thought that was a highly revealing remark. It is a common misunderstanding in science that there is only one truth, and that to each question there is only one true answer. Lewis Wolpert, developmental biologist and great defender of scientific inquiry, has often expressed this idea of univocal truth openly. The idea contains two misunderstandings.

First, all scientific ideas are hypotheses, not dogmas, and as such should be open to question and falsification. It was a great mistake, for example, for Francis Crick to categorize the one-way coding relationship between DNA and proteins as the "central dogma," as though the "one truth" had been found. As University of Chicago bacterial geneticist Jim Shapiro shows convincingly in *Evolution: A View from the 21st Century*, today there is not much left of the central dogma other than a chemical fact about DNA forming templates for proteins, whereas proteins don't reciprocate by forming templates for DNA. Surprisingly, perhaps, the deconstruction of the central dogma has come from within the citadel of molecular biology itself. Control signals are continually traveling from the organism to its genome, and the effects during evolution include large-scale reorganizations of the genome, with whole sequence domains moving from one gene to another. As P. J. Beurton and his colleagues conclude in *The Concept of the Gene in Development and Evolution*, "It seems that a cell's enzymes are capable of actively manipulating DNA to do this or that. A genome consists largely of semistable genetic elements that may be rearranged or even moved around in the genome, thus modifying the information content of DNA."

Second, that kind of answer, and the related one that philosophy is not necessary to science, fails to address the problem. Some scientific questions cannot be formulated coherently without adopting a point of view that already implies deep and essentially philosophical commitments. Perhaps the best examples of that in modern science are to be found in physics, particularly in areas like quantum mechanics and cosmology. Both require speculation of a kind that is best described as metaphysical. Indeed, many theoreticians of quantum mechanics openly admit that their equations are not there to describe reality "as it really is," whatever that statement might mean. The physicist and mathematician Henri Poincaré, who anticipated many, if not all, of the ideas of relativity theory, wrote a beautiful book, *La science et l'hypothèse*, outlining the essentially philosophical nature of much theoretical work in science. He also makes the telling point that the worst mistakes are made by those who claim that they are not philosophers. They not only fall into the holes but also don't even recognize that there are holes to fall into!

My experience in writing *The Music of Life* had already led me to encounter several ways in which the development of evolutionary theory runs into these kinds of problems, which are often subtle and involve what can best be described as linguistic and philosophical confusions. Samir Okasha's book *Evolution and the Levels of Selection* is a powerful and careful corrective, since he brings philosophical rigor to the debates.

So I was already prepared for the next encounter with Lynn Margulis. Her mission then became clearer when she repeated the question but this time with the emphasis changed: "You *are* Denis Noble?" She had clearly read my book *The Music of Life* and wished to record me and other Oxford biologists, such as Martin Brasier, talking about evolution. This was when I first met Jim MacAllister, whose full role in this story will become evident later. He set up the recording equipment in the beautiful Bajpai room of Balliol College. I had agreed to bring my guitar and to do what I often do in popular lectures on *The Music of Life*, which is to illustrate the ideas with performed music. I used an Occitan love song to introduce the story of the genome as a CD, which I describe in chapter 1 of the book. The idea is simple but persuasive. Genes, as DNA sequences, are read by cells and organisms just as a CD player reads the digital data on the disc. Both are best viewed as databases. It would be a mistake to identify the digital data on a CD as the music itself. The same kind of mistake is involved in describing the genome as the "book of life."

Lynn warmed to the way in which I used competing metaphors to illustrate the ideas of systems biology and how those ideas were challenging some of the central assumptions of twentieth-century biology. The recording therefore seemed to be what she wanted. A question she then put to me left me floundering. "Could you comment on the idea that evolution had come about through the gradual accumulation of random mutations?" I had no idea! At that time I had not read the work of Jim Shapiro and others showing how far from random mutations are. I was still stuck with what I had been taught on this question as a student many years ago. Worse still, at that time I had not appreciated how critical this question is to whether physiology is relevant to evolution.

In fact, it is central to that question. If mutations are random, then they are not guided by any physiological property of the organism or by that organism's interaction with the environment. But if they are not random, and if the changes are in some way guided by functionality, then the whole game changes. Physiology becomes the means by which we can understand how the organism reacts to its environment and communicates this to the genome. That idea is not new; Charles Darwin also had the essence of the idea. Puzzled himself by some of the phenomena he also interpreted as the inheritance of acquired characteristics (there are twelve such passages in *The Origin of Species*), he invented the idea of gemmules, particles that he thought might be transmitted from the tissues and organs of the body via the bloodstream to the germline cells. You and I carry such "particles" in our bodies from even before our birth: maternal cells transmitted via the bloodstream through the placenta, responsible for some of the so-called maternal effects.

The second part of Lynn's mission became clear when she announced to me that she had persuaded Richard Dawkins to take part in a debate with her, that she was arranging this to happen in Balliol, and—wait for it!—she "commanded" me to chair it. By now, I knew Lynn well enough to know that there was no way in which she was going to accept a refusal to do what she had commanded.

The debate took place in May 2009, and it was an extraordinary event, recorded in full by *Voices from Oxford* (for video) and Jim MacAllister (for audio). It can be viewed and heard on the *Voices from Oxford* website. (Please see page 186 for website URL.) It can be seen as Balliol's tribute to her memory. She introduces the debate with "Welcome to Balliol [emphasis!]—the best place I know."

First of all, though, let me say that, although I take a different view of biology from Richard, I admire his writing skill, his ease of communication with the public, evident in many of his books and broadcasts, and I think we owe him a big vote of thanks for so generously taking part in the debate, through all four hours that it lasted. He referred in his opening remarks to neo-Darwinism as a theory that "had taken some stick" recently. No one could doubt that he had Lynn in mind. He also anticipated—or tried to defuse—some of Lynn's remarks by saying to Lynn: "I don't believe you go far enough!" (By which he meant genes were themselves "symbiotic," joining up with other genes in other bodies.) Lynn and Richard had of course exchanged some pretty rough words about each other's theories in various public pronouncements. Some of those remarks are almost unprintable. I was therefore a little concerned about how the debate might be chaired. In the event, I need not have worried.

Lynn had also recruited two other Oxford specialists in early-stage evolutionary events: Stephen Bell on microbiology and Martin Brasier on the pre-Cambrian fossils. Together with Lynn's great film on symbiogenesis, there was a lot of valuable new material in the presentations during the early stages of the debate. The audience, too, contained some high-powered people, some of them from the seminar series that have now become part of the Balliol Interdisciplinary Institute (BII). Controversy is there in the debate, and the people concerned, both from the audience and the debaters, did not pull punches. It was a unique event. With Lynn sadly gone, a debate of this sort can't ever be repeated.

In Lynn, we have lost an irreplaceable heretic. As such, she rapidly discovered which heresies I was responsible for. The danger now is that those heresies are well on the way to becoming orthodox. I say that because book after book, article after article now conveys the message that, at the least, the modern synthesis needs extensive revision, perhaps even replacement by a new synthesis. As that process continues in future years, we will have Lynn to thank for a major push in that direction. If she was harsh herself and even nondiplomatic in her own comments on neo-Darwinism, I think that was partly the reaction of someone who had got tired of the way in which slavish adherence to the theory was blocking thoughts and experiments that needed to be thought and done. There comes a time when orthodoxy should recognize the power of the challenge it faces and adapt or give way gracefully to the new ideas.

Denis Noble of Balliol College, Oxford University, is a physiologist who contributed to our understanding of the chaotic dynamics of heartbeats. A student of medieval music, he is author of The Music of Life.

Neo-Darwinism

and the Group Selection Controversy

JOSH MITTELDORF

I first met Lynn Margulis in her UMass office one day in July 2006. She dismissed me as a neo-Darwinist—the worst insult in her vocabulary—and I was devastated.

Lynn's life spanned a golden age of biochemistry and a dark age of evolutionary theory. She saw the tools of microbiology come into their own—the electron microscope became a basic laboratory tool; the metabolic chemistry of the cell was deconstructed; and computer analysis of genome sequences became routine. It was also a period of time in which evolution became dogmatic and imperious, when bench scientists and geneticists learned to accept a rigid framework for understanding the evolutionary provenance of what they were seeing, whether or not the theory fit the observations. The theory in question was a narrow and parochial interpretation of Darwin's theory, sometimes called population genetics, also known as the modern synthesis or neo-Darwinism.

For half of the twentieth century, evolutionary biology had been two separate disciplines, developing on parallel tracks. There were the naturalists, traveling and collecting stories about animals and plants, pronouncing loosely on the evolutionary provenance of the traits and features they observed. And there were the mathematicians, developing a rigorous, quantitative theory for the operation of natural selection. The two schools spoke different languages, deployed different concepts, understood the world through different modalities. The naturalists were intimidated by mathematics. The mathematicians were uninterested in the messy details of real biology and preferred to see how far they could extend their theories based on logic alone.

Had science progressed according to the way things are supposed to work, there would have come a point when the mathematical theorists compared their predictions to the world that was being described by the naturalists. Biological reality would have served as the ultimate arbiter of theory, and the theorists would have adjusted their equations accordingly. But that's not what happened. Right around the time that Lynn was beginning her career, the theorists were executing a hostile takeover of the field of evolutionary biology, and Darwin's broad, open-ended theory of evolution was acquiring the straitjacket of neo-Darwinism. The lines were formed, and the battle was joined over the battlefield of group selection.

On its face, the group selection controversy was about whether natural selection operates strictly among individuals, or whether it may sometimes work among groups and communities. In many ways it doesn't matter, because the qualities selected would be the same in any case. For example, fast gazelles might have an advantage over slow gazelles in escaping from predators. You can think of a fast herd of gazelles, or you can think of fast individual gazelles in a herd, and either way you come to the conclusion that evolution selects for speed. But sometimes the interest of the individual and the interest of the group are in conflict. Suppose a sentinel meerkat stands guard watching for danger while the rest of the pack forages for food and whistling if he sees a predator. The sentinel doesn't get enough to eat and exposes himself to extra danger by standing tall. If we think in terms of individual selection, then he is likely to be less successful than other meerkats in the group in passing on his genes, and this behavior could never evolve. But if we think in terms of group selection, then a meerkat group with a sentinel is likely to be more successful than other groups, so we would predict that this behavior offers an evolutionary advantage and should be selected.

The neo-Darwinists analyze this situation in terms of "kin selection": if the meerkats are in a family group with closely related individuals, then the sentinel meerkat is probably protecting others that carry the same gene for sentinel behavior; but if the group members are unrelated, then it's more likely that the sentinel is protecting others that don't share the sentinel gene. So according to the neo-Darwinists, the behavior should evolve only for close family groups. But group selectionists would admit another possibility: the sentinel behavior provides a benefit to the group whether or not the meerkats are closely related, and if groups of meerkats with a sentinel are able to survive better, then in the long run the behavior might evolve nevertheless, perhaps via an evolutionary process that is complex and difficult to analyze. In this case, the salient fact is that meerkat "mobs" (as they are called) are not always closely related.

The simple, scientific question is whether natural selection operates strictly one gene at a time, or does evolution also select for communities and groups of unrelated individuals? But scientists are human, and this question acquired a great deal of cultural baggage that biased their thinking in one direction or the other. Individual selection is associated with social Darwinism and legitimacy of social strata based on individual economic success. Group selection is associated with a caring and paternal social philosophy, sympathetic to the social safety net and the welfare state. Even the language of science has overtones that load the question with emotional and political significance. The sentinel's behavior is called "altruism," and the family-group hypothesis is called the "selfish gene."

Ronald A. Fisher, working in the early twentieth century, was the founding father of neo-Darwinism. Fisher was a mathematical genius and single-handedly devised many of the techniques that underlie the statistical analysis of data in every science today. Fisher was also an outspoken racist and passionate advocate for eugenics. He feared that it was the "least fit" members of society who were outbreeding the aristocrats and threatening to dilute the mental resources that make human civilization possible. To this day, neo-Darwinists tend to be free-market libertarians, while the subcommunity of evolutionary biologists who believe in group selection tends to the political Left. The period from 1970 to 1985, when the selfish gene was being enthroned as the only viable mode in which Darwinian selection operates, was just coincidentally the same era in which the unrestrained competition of free-market capitalism was hailed as the one true economic system.

V. C. Wynne-Edwards was an exemplary practitioner of the old school of naturalism. In 1962 he published a book that culminated a life's work on biological communities. He described, citing dozens of examples, the ways in which animal populations restrain their reproduction to avoid overpopulation and safeguard the ecological resources on which they collectively depend. A few years later, a smart young biologist named George Williams published his rejoinder, in which he took Wynne-Edwards to task for lack of rigor in his theoretical reasoning. Wynne-Edwards was implicitly invoking group selection, and Williams doubted that this process played a role in Darwinian evolution. It was not theoretically possible, Williams wrote, that natural selection could operate in the way that Wynne-Edwards claimed.

A debate ensued over the following decade, unfolding in pages of evolutionary journals that had once been filled with observations and descriptions but now tended increasingly toward sophisticated mathematics. The theorists were more domineering and clever. Intimidated by mathematics and burdened by a fuller appreciation of the complexity and ambiguity of Mother Nature, the old-school naturalists were no match for them. The only thing they had going for them was the facts, as I explain below.

Laboratory evolution had a role to play in this debate, but it was not as constructive as scientists had hoped. Breeders of fruit flies and roundworms learned to isolate a trait of their choosing—the color of a fly's eyes or the propensity of a worm to swim toward food—and select the trait "by hand" in a lab population. Lo and behold, the breeding worked, and they found the trait to be enhanced in each succeeding generation. This was exactly the way the neo-Darwinian process was supposed to work and was taken as experimental corroboration of the theory: selfish-gene theory was found to work in the laboratory just as it works on the blackboard. This limited success tended to mask the larger question: is this how nature works?

The theorists won. Altruism was declared to be an illusion. It was decreed that any adaptation that *appeared to be* beneficial to the community must have an alternative explanation from the perspective of the selfish gene. A wave of theoretical research followed in which an author would cite an instance of apparent altruism and propose a mechanism by which it *might have* evolved through selfishness. Two generations of university students were instructed that "this is the way in which evolution works."

Neo-Darwinism recognizes just one mode of natural selection: compare a population before and after a single generation of a single species. The world stays the same. The ecological background remains the same. The population remains the same size. But individual genes increase or decrease in their prevalence within that population, based on their average contribution to the fitness of their owners. So "fitness" in neo-Darwinism is simply a matter of how many offspring an individual contributes to the next generation. It is, by postulate, the "target of natural selection," meaning that this is what evolution works to maximize.

This paradigm was adopted by Fisher because it is simple enough to admit neat and elegant mathematical analysis.

Reality is a different story, or not a story at all. It consists of phenomena including some that are messier and more interesting than those modeled in Fisher's equations: overpopulation, leading to crashes and local extinction; cataclysmic extinctions, in which whole ecosystems collapse because of meteor impacts or geographic or climate changes; coevolution, involving symbiotic negotiation across species lines; horizontal gene transfer, in which DNA migrates through viruses or bacteria into the genome of a "higher" plant or animal; and Lynn's trademark, endosymbiosis: twosomes, threesomes, foursomes, and moresomes of organisms that merge their bodies and their genomes to form new, more complex entities.

Neo-Darwinists don't deny that these things occur, but they push them to the edge of their science, treating them as exceptions. Such things happen, they aver, but they're not relevant to the way that evolution works in the main. Meanwhile, there are major aspects of the biosphere that by their nature defy explanation in terms of selfish genes. Three of these may illustrate ways in which the narrow framework of neo-Darwinism is challenged beyond its limits: sex, aging, and hox genes.

Sex is so closely identified with reproduction that we forget that once upon a time sex and reproduction were quite unrelated. Sex is the sharing and mingling of genes. In protozoans ("protists" is the word biologists use), reproduction happens by simple cell division (mitosis), whereas sex is a separate, less frequent process (conjugation) in which two cells come together, merge their cell nuclei and their genes, and then separate as two hybridized individuals. From the perspective of the selfish gene, sex is a terrible idea. It breaks up combinations of genes that have a track record of working well together and casts the dice with a combination that's new and unproven. If my genes are working well, the last thing I want to do is share them with my rival.

For higher animals that have male and female forms, sex is an even worse deal. Fully half the population is unable to reproduce! Darwin says that the race goes to "the fittest," and neo-Darwinists measure fitness by the number of offspring a plant or animal produces (and the speed with which they are created). By this standard, we would all be twice as fit if we were hermaphrodites, like snails or flowers—able both to fertilize others' eggs and to produce eggs of our own.

In fact, individual selection against sex is so strong that sex could not be maintained without being tightly linked to reproduction. We're all familiar with the fact that for higher animals and plants the very machinery of reproduction involves sex in an essential way. But even in protists the imperative to share genes is built into the life cycle: a mechanism called "replicative senescence" ensures that any cell lineage that *doesn't* submit to conjugation dies out after a few hundred generations.

Aging, too, is a bust from the point of view of the selfish gene. The theorists have decreed that it could *not* have evolved as an adaptation via a process of natural selection. An aging gene is just the opposite of a selfish gene, and what it offers is the opposite of fitness. If aging is ubiquitous in nature, it must be because natural selection doesn't care very much. Few animals and plants in nature ever live to be old enough for aging to matter (decreed the theorists), so advanced ages are "invisible" to natural selection, and aging, with no "adaptive" or functional purpose, has been permitted to degrade through the natural selection equivalent of neglect.

In the 1980s and 1990s, field studies were designed to answer the question: Is it really true that there are no old animals (or very few) in the wild? Or does aging turn out to be a significant drag on individual fitness? The answer was a surprise to everyone: aging takes a huge bite out of fitness. Flies and tropical animals may be losing one-fourth of their potential offspring because of death from old age, so in the language of individual fitness, aging has stolen 25 percent of their fitness. For larger mammals in the Arctic, this figure can be as high as 80 percent. This is not the way things were supposed to work.

And then the bombshell: as molecular genetics matured in the 1990s, genes for aging were found. Remove the aging gene (via gene splicing or RNA interference), and the animal lives longer! Many such genes were discovered, and, what's more, these genes are ancient—preserved over a billion years of evolution, their descendants identifiable in the genomes of all manner of living things, from worms to mammals. Why would nature be preserving these genes if they "don't matter"? Worse still, if these genes are really so bad for fitness, why does natural selection seem to be guarding them like crown jewels? This is a powerful indication that nature's definition of "fitness" is not the same as the neo-Darwinists' definition. Genes for aging are prima facie evidence for the phenomenon, if not yet a full mathematical theory, of group selection.

In fact, over the eons, not only has the genome learned how to survive and reproduce; the genome has learned how to learn. Chromosomes have been structured hierarchically, with control genes—hox genes—that act like master switches, turning on whole developmental programs or creating entire organs. A single hox gene can give the command "make an eye here," and all the tissues and constituent parts will develop in response. This enables evolution to experiment with the number and placement of eyes, without risk of destroying the anatomy that has been so highly optimized. Hox genes are the genetic basis for development of the body, and they are as old as multicellular life—suggesting that they played an essential role in the transition from one-celled forms half a billion years ago.

Structuring the genome with hox genes works over the long term. The use of hox genes offers a great advantage in that it makes the process of evolution far more efficient and flexible. This is not at all the kind of individual advantage that can be expressed in terms of selfish genes. (The individual with hox genes doesn't survive better and produces no more offspring than the individual without hox genes.) Rearranging the genome with command-and-control logic must have required a great number of generations, during which time genes spread through large populations. This is group selection on a grand scale. The very existence of hox genes is impossible to understand within the framework of neo-Darwinism; and besides hox genes there are other equally mysterious ways in which chromosomes seem to be optimized, not for fitness but for the process of evolution itself. Natural selection seems to work not only for traits that contribute directly to survival and rapid reproduction but also for traits that help whole evolving populations over extended periods of time.

Throughout her career, Lynn bore witness against the tenacity of theory in the face of countervening facts. In *Acquiring Genomes* she wrote, "No adequate quantitative measure of group selection [benefit, higher organism, inclusive fitness, and many other terms bandied about in neo-Darwinism] exists. . . . Therefore these are deficient, even pseudo-scientific terms." In *Symbiosis in Cell Evolution*, she writes, "Different bacteria form consortia that, under ecological pressures, associate and undergo metabolic and genetic change such that their tightly integrated communities result in individuality at a more complex level of organization." What we see here is not an engagement with but a very rejection of the terms of the group-selection debate. She wrote with a deep understanding of the incontrovertible evolutionary datum, studiously ignored by neo-Darwinian zoologists who specialize, if at all, in only a very few animal species, that animals themselves are the result of group selection among colonies of archaea, bacteria, and eukaryotic cells. In short, except when she was feeling argumentative about the neo-Darwinian coup and its fondness for mathematical theory with woefully few data points in the real world of natural history so beloved by her and Darwin, she dismissed the very terms of the debate in favor of a grander, more naturalistically grounded vision, as was her style. Her instinct was always to work upward toward theory from the facts on the ground, rather than explore consequences of theory from the top down. The facts are that the biological world is much more interdependent, better integrated, and more tightly coordinated than even the most starry-eyed group selectionist can account for. Based on the reality of these observations rather than any theory, Lynn left behind the squabbles about whether group selection is absolutely excluded by theory or whether it may sometimes compete with individual selection. She embraced radical ideas of group adaptation on larger and larger scales—right up to the level of Gaia, the biological community of Mother Earth, which, as far as we know, is not engaged in a competitive struggle for existence with any other blue planets.

> Evolution favors populations of individuals that act together to re-create individuality at ever higher levels. This somewhat freaky assertion calls into question the very usefulness of trying to isolate the units of natural selection: because

> of the articulation or community relations of living things, the differential reproduction of units at one level translates into the differential reproduction of units at a higher, more inclusive level.[1]

This perspective is only now beginning to gain adherents within the community of evolutionary biologists, where the phrase "group selection" is still a hot button.

Josh Mitteldorf uses mathematical modeling for research in evolution. He holds a PhD in theoretical astrophysics from the University of Pennsylvania and has published articles on the genetics of group selection and aging in Evolution, Science, *and* The Journal of Theoretical Biology.

A Modern-Day Copernicus

Sippewissett Time Slip

STEFAN HELMREICH

The salt marshes of Sippewissett, Massachusetts, a few miles north of Woods Hole's Marine Biological Laboratory, host sheets of microbes that detail a history of early Earth. This history is not so much archaeological—revealed by peeling back ever-older layers—as it is analogical: microbial mats similar to these have likely existed on Earth for more than three billion years.

I visited these squishy structures in May 2005 with Lynn Margulis, traipsing with her through the Cape Cod mushscape as part of anthropological fieldwork I was then doing about how contemporary microbiologists have been reimagining the past, present, and future of our ocean planet. I was conducting ethnographic fieldwork, it turns out, about, alongside, and entangled with Margulis's microbiological fieldwork.

The ocean I was in the midst of discerning was one that I have come to call the "alien ocean." That ocean, for today's marine microbiologists, manifests as both a futuristic, science fictiony space of weird and ultimate others—critters living at extremes of temperature, pressure, salinity, and much else—as well as an alluring reminder of our own organismic origins in brine and bacteria. To adapt an old anthropological dualism about the simultaneous newness and everydayness of "other cultures," today's microbial sea is both strange and familiar. It is an echo of the past as well as a place unfolding into unknown futures.

"Strange and familiar" well describes the wet and watery world that Margulis described in her work. She was in Sippewisset pondering additions to her theory of symbiogenesis, the idea that evolutionary novelty emerges from the symbiotic fusion of different sorts of cells and organisms. According to this view, all of today's nucleated, eukaryotic cells evolved through incorporating once free-living prokaryotes—like the oxygen-respiring bacteria that became mitochondria and the cyanobacteria that became chloroplasts, entities that now constitute indispensable organelles in the cells of animals and plants. In Margulis's tale of symbiogenesis, the biologically strange has become the biologically familiar—the familial, even.

As we walked through the marsh, Margulis told me that she was investigating the possibility that structures such as the tiny hairs on the edges of cells, the filamentous threads in mitotic cell division, and the tails of sperm—all known to have common descent—might come from an earlier incorporated ancestor called a spirochete, a swimming, corkscrew-shaped bacterium. Margulis was leading us to the marsh to collect mats of organisms she knew hosted similar spirals stirring from a dormant, rolled-up form; she wanted to detect signs of emergent symbiogenesis, evidence that these wiggly creatures were insinuating themselves into their neighbors' cellular structures.

There were seven of us on this modest expedition into Earth's microbial past: Margulis, some students from Woods Hole, a journalist from *Discover*, my anthropologist spouse, and me. We trailed Margulis's red compact car over dirt roads that string together gray-shingled Cape Cod vacation homes along the marshlands. Private Property and No Trespassing signs did not deter Margulis; she moved in a different time and space continuum than most New England humans. As a key collaborator with James Lovelock on Gaia since the 1970s, her mind tuned in at once to the microscopic and macroscopic, to the subvisible and superorganismic, to the baroque and romantic aspects of the world around us. She tuned in to what we could call a Sippewissett time slip, playing here on the title of Philip K. Dick's *Martian Time-Slip*, a novel that takes place on a Mars that is experiencing all its history simultaneously. Our Sippewissett trip takes place not just in 2005 but also in another time, on an early Earth, in a throwback sea. One might note, too, that "Sippewissett"—a Wampanoag place-name meaning "at the little river"—also indexes the all too ghostly Native American presence in this patch of Cape Cod. This place is a mix of past and present: a time machine, an eddy in the alien ocean.

After parking our caravan of cars—fueled by oil, itself the result of ancient microbial processes Margulis calls "unearned resources"—we hiked out to the intertidal zone, the sort of threshold region Rachel Carson celebrated in her ecological urtext from 1955, *The Edge of the Sea*. Margulis wore knee-length waders and a fleecy hooded pullover and walked determinedly into the mush, grasping a spatula and large spoon, ready to hunt microbes. She was the only one dressed appropriately, though sympathetic to those of us freezing in the drizzle on that unseasonably cold day. Waving us into the marsh, she pointed out grasses that were signs of nearby microbial mats.

She directed us to shallow pools bottomed by multicolored mud. Pointing out that the alternately exposed and submerged character of this site is ideal for mat growth, she dug up a lichen-like sample of microbial mat and instructed us to look at its dripping wet cross section, half an inch thick. Its rainbow layer cake of orange, green, pink, and black was composed of diatoms, cyanobacteria, purple sulfur bacteria, sulfate-reducing anaerobes, and other microbes living in complex interdependency. The mats were ecosystems, with cyanobacteria and purple anoxygenic photosynthetic bacteria as primary producers—"the acme of evolution," she remarked.

"Cyanobacteria can do everything," Margulis said, "except talk." "Everything" means they gather their necessaries from vastly different media: electrons from water, energy from sunlight, and carbon dioxide from the air. They are expertly suited to the earthly elements, the proportions of which, according to the Gaian model, they have themselves had a role in determining.

Margulis was inviting us to look anew at planet Earth, to peer through these mat systems all the way back to the Archean eon, 2,500 million years ago. The mats were portals to another world. They were media transmitting messages from an age when the ocean was otherwise, when Earth was just becoming the planet it is now. Cyanobacterial photosynthesis long ago filled the world with oxygen, pushing into marginal zones such anaerobic creatures as methanotrophs. Ancestors of blue-green bacteria like the ones in microbial mats consigned methanotrophs to environments we now consider extreme—the reason scientists call them and other such nonstandard life-forms extremophiles.

Standing in the marsh with Margulis, we were searching for embodied, living analogues of ancient life. There is ample evidence for mats in Earth's fossil record in the form of stromatolites, rocks left behind by the trapping, binding, and deposition of carbonate (and other sediment, such as sand) by cyanobacterial mats. Some biologists suggest that if bacteria evolved on other planets they would leave behind stromatolite-like formations, which might then be considered signatures of extraterrestrial life. The extraterrestrial analogy was not far from our minds because our trip into the marsh came at the end of a weeklong workshop on astrobiology at Woods Hole's Marine Biological Laboratory. Astrobiology is an area of inquiry devoted to thinking about biological systems—actual ones on Earth and possible ones elsewhere—in a cosmic context.

Astrobiologists want to know how life emerges on worlds in general and in particular; they are curious, for example, whether Mars might host microorganisms akin to those found in extreme environments on Earth. In this scientific venture, microbes in such locations as salt marshes become meaningful as proxies for extraterrestrial life. Scientists read them not just for clues about Earth's past, present, and future but as a means for considering the category of life itself in a more ample, universalistic frame. Margulis's work on comprehending earthly life as transformative of planetary biogeochemistry has been centrally important to astrobiology—a field of inquiry institutionally established by NASA in the late 1990s though in existence in a slightly different version since 1960 as "exobiology."

At the astrobiology workshop, Margulis gave us a compressed, autobiographically organized chronicle of the Gaia hypothesis, which she joined Lovelock in developing in 1970. As she told us, this model originally emerged to look not at Earth but at Mars, to determine whether it might be possible to discern from a distance whether the red planet supported life. In 1965 Lovelock had been invited by NASA to design an experiment for detecting life on Mars. Using his expertise in gas chromatography, he wagered that the best way to look for life remotely would be to search for signs of metabolism in planetary atmospheres. As our workshop organizers, historians Steven Dick and James Strick, put it, Lovelock suggested that "the most obvious activity of living things which offsets entropy [is] that they keep the gas composition of a planetary atmosphere far from equilibrium."[1]

Margulis emphasized that the Gaia hypothesis did not suggest that the planet was some perfect Eden, as many critics have misunderstood it to claim. She pointed this out during our jaunt into the salt marsh: "The ocean is too salty. Does anyone know the pH of the ocean? Most biologists will say 7 because that's neutral and they want to be neutral. But it's not. It's 8 or so. The ocean is too salty." Fishes are often happier, she remarked, in the lower salt concentrations of water provided in aquariums. "Gaia is not God and didn't do anything perfectly." Margulis emphasized that Gaia could care less about humans. She dismissed, as well, the idea that Gaia demanded that Earth be considered an organism: "No organism can consistently eat and live on its own waste." If Gaia sometimes veers toward the romantic—a holistic vision of harmony—it also includes an attention to such baroque complexity as the never fully equilibrated relation between chemistry and biology.

How far can Earth or life be translated into theoretical terms that can float free of particular embodiments? If we follow Margulis on her travels in the salt marsh, the answer might be "not too far." Margulis periodically toted mud from Sippewissett back to her lab in Amherst, where she placed mats in nutritive media to see if spirochetes would materialize and attach themselves to other cells. As she wrote in 2004, "I believe that with much help from colleagues and students, we will soon be able to show that certain free-swimming spirochetes contributed their lithe, snaky, sneaky bodies to become both the ubiquitous mitotic apparatus and the familiar cilia of all cells that make such 'moving hairs.'"[2] For Margulis, living things were—are—forever incorporating one another, engaging not just in lateral gene transfer but also in lateral genome transfer.

But Margulis would not know how such incorporation works unless she actually ran an experiment, tried to jostle spirochetes awake to see what they might do next. In the language of rhetorician Richard Doyle, she worked in the realm of "wetwares," "encounter[s] with flesh as a refrain, a repetition of algorithms or recipes of sufficient complexity that only through instantiation can they be experienced."[3] In other words, Margulis's spirochetes produce a sign of life that needs to be *fed* to be *read*. The semiotics of life needs living things to signify, and those things cannot exist except in contingent, real time—coming into liveliness through such material activities as eating, which always happens in a web whose coordinates are never fully in place prior to their habitation and creation. "Life" is a set of relations of sustenance, operating across scales. There is no Platonic world of "life."

Margulis's biology, then, is a fully theoretical biology that does not permit theory to operate as an abstraction, to rove over or above the bodied enfleshments that are living things. In her 1995 book with Dorion Sagan, she addressed the question "What is life?" by delivering a distinct answer to the question for each of life's five kingdoms: bacteria, protoctists, animals, fungi, and plants—emphasizing neither some underlying logic nor an overarching metaphysics but rather the situated particulars of bacterial, protoctist, fungal, plant, and animal embodiment. Life was not something that could be compressed into the logic of a code but was a process ever overcoming itself in an assortment of bodied manifestations.

Thinking back on Lynn Margulis's leading of our crew into the Sippewissett time slip, then, I might put the lesson this way: life, like the sea, is an alien, a visitor always on its way toward fields of unexpected connection.

Stefan Helmreich, an anthropologist at MIT, is author of Alien Ocean: Anthropological Voyages in Microbial Seas.

The Cultural Dimensions

of Lynn Margulis's Science

WILLIAM IRWIN THOMPSON

When twentieth-century genetics was added to nineteenth-century natural selection, we were given a grand narrative of how the competition of individuals for survival and reproduction created an ecosystemic marketplace—one disturbed from time to time by catastrophes that knocked out a gene in random mutations caused by an errant cosmic ray or a falling star. A bolide or a supervolcano that darkened the atmosphere for years was presented as a rocky speed bump on the path of the evolution of the biosphere that could cause species within it to crash and become extinct.

Lynn challenged both these ideas of mutation and catastrophe. Mutations, she claimed, generally cause more damage than naturally selectable novelty, and catastrophes, like a horizon, are based on the observer's perspective. The oxygenated atmosphere produced by photosynthesizing cyanobacteria was not a catastrophe for bacteria but simply a change of address. The bacteria are still here, in our guts and at the bottom of lakes.

Or, to speak in the metaphors of my native Celtic animism that Lynn appreciated in my descriptions of her critters as "the little people," the dwarves like to work underground in the mines, but the elves prefer to move in the air. (Notice in Tolkien's *Lord of the Rings* that the dwarves prefer iron hammers, and the elves prefer arrows.)

Symbiotic transformation, not competition, is the central trope in Lynn's narrative, so much so that acquired genomes are more important than mutation in the architecture of life on this planet. Lynn's narrative was no longer simply a footnote to Darwin but a new grand narrative. Postmodernist Jean-François Lyotard got it wrong: the great age of grand narratives, like those of Darwin, Marx, and Freud, was not over. *Les petits récits* of postmodernism were simply the little notes and billets-doux that academics liked to send one another back and forth in their specialized journals. Lynn was no trendy postmodernist; she was a big thinker, and the academics and funding bureaucrats didn't like her for it. She never received the major National Science Foundation funding, a MacArthur, or the Nobel Prize that she deserved.

Science, as Ludwik Fleck taught us in his 1935 classic *Genesis and Development of a Scientific Fact*, which Lynn turned me onto, is a social process, and a fact cannot exist without a theory anymore than a candle flame can exist without an atmosphere. A theory is a narrative, and a narrative that answers the three questions of Who are we? Where do we come from? Where are we going? is a narrative that amounts to a myth.

If you think in the neocon terms of competition in a marketplace as the nature of life and society, then you are thinking mythically. If you think identity is packaged in a gene that has "simple location" in the nucleus, you are also thinking mythically. Lynn knew that identity was a process and not a location.

To understand Lynn is to understand that her narrative has profound political implications, just as Darwin's did. Consider the following passage from *The Origin of Species*:

> In each well-stocked country natural selection acts through the competition of the inhabitants, and consequently leads to success in the battle for life, only in accordance with the standard of that particular country. Hence the inhabitants of one country, generally the smaller one, often yield to the inhabitants of another and generally the larger country. For in the larger country there will have existed

> more individuals and more diversified forms, and the competition will have been severer, and thus the standard of perfection will have been rendered higher.

In the sociology of knowledge, what is being presented here is not simply a description of the domestic breeding of pigeons but the history of England and Ireland in the nineteenth century. The mentality expressed here is the worldview of imperial British capitalism, in which competition, the battle for life, and the telos of evolution lead to higher standards of perfection.

In other words, it is not the case that we have successive historical waves of Darwinism, and then social Darwinism. Social Darwinism is already present in Darwinism.

What is true of Darwin is also true of Margulis. If identity is a process and not an object like a gene, then the process and reality of symbiosis and acquired genomes lead to new narratives of cultural cooperation and mutual interpenetration in space and time.

If we understood the political implications of Margulis's grand narrative, then our worldview would generate a radically different phase-space for Israel and Palestine or Northern and southern Ireland. We would not seek to apply a fixed identity to a "simple location."

Had Americans not been so taken with racist visions of development in the nineteenth century, the history of our country could have been different, as Jefferson and Tecumseh tried to work out a more complex cultural development of the Louisiana Purchase than the enforced expansion of Manifest Destiny and Empire, resulting in the Oklahoma Sooners and the Cherokee Trail of Tears.

In acquiring genomes it is not just the melting of membranes and the creation of aggregates but the proximate association in which the art of the dominated influences the culture of dominating. Irish literature in English paved the way for the Anglo-Irish Treaty. Before you can have a President Obama, you first need to have a Duke Ellington.

If you believe that the cell is an information-processing machine for reading the genetic code, if you think evolution is a struggle of discrete units to survive long enough to reproduce, if you think the human world, like Maxwell's demon, is a difference engine for sorting out creatures in a competitive marketplace of rational self-interest, if you think the immune system is a military force defending self against other, or if you think the nation-state is threatened by the infection of the other in the form of alien immigrants, then you need to study the works of Lynn Margulis, because everything you think is wrong. Lynn, like Darwin before her, has changed everything.

William Irwin Thompson is a cultural historian, poet, and founder of the Lindisfarne Association.

Lynn Margulis on Spirituality and Process Philosophy

DAVID RAY GRIFFIN AND
JOHN B. COBB JR.

Although we had for many years admired Lynn Margulis, especially the courage and evidence with which she challenged neo-Darwinism, our relationship with her began in 2004, when she participated in a conference on evolutionary theory that one of us (Cobb) had organized.[1] Besides quickly becoming friends, we came to see—during the conference and then later through additional reading and conversation—that we had even more in common than we had initially realized.

In this essay, we point to ways in which we, in spite of having had very different backgrounds and educations, ended up with similar visions. Because of the similarities of these visions, we are able to use her empirical discoveries in support of our philosophical position. And insofar as this philosophical position, generally called "process philosophy," is found to exemplify the standard signs of truth—consistency, adequacy to experience, and illuminating power—this position can be used to undergird Lynn Margulis's scientific vision.

Because we are theologians as well as philosophers, we were asked to address the question of her spirituality. For most of her career, religious and spiritual matters, as usually understood, were evidently not high on her conscious agenda. Indeed, she explicitly rejected "belief in God" and other conventional signs of religiosity. More important than such signs, however, is whether a person is deeply committed to truth, beauty, and goodness. Viewed from this perspective, Lynn's entire life reflected a deep spirituality. Moreover, because of her commitment, especially to truth, she made quite remarkable contributions to the spiritual quest of others, especially after she began supporting—and enriching—the perspective that James Lovelock had come to call "Gaia."

In this brief statement, we simply name a number of ways in which Margulis's worldview is consonant with that of process philosophy. This philosophy is based primarily on the writings of Alfred North Whitehead, who, having originally focused on mathematical physics and logic, came to develop—after moving from England to Harvard in the 1920s—the most extensive philosophical system of the twentieth century, which he called "the philosophy of organism."

Organism

Whitehead developed "a system of thought basing nature upon the concept of organism," which he contrasted with the concept of mechanism.[2] "Science," he said in 1925, "is becoming the study of organisms. Biology is the study of the larger organisms; whereas physics is the study of the smaller organisms."[3] In referring to a cell as an organism, he meant (1) that it experiences other organisms (although "experience" does not necessarily entail consciousness), and (2) that it has a degree of spontaneity. Although Whitehead referred to some enduring entities—such as cells, molecules, atoms, and electrons—as organisms,[4] he thought of organisms in the most fundamental sense as momentary events, which he sometimes called "drops of experience."[5]

Margulis, properly restricting herself to biology, did not deal with the entities of physics. But she described the basic biological entities in organismic, rather than mechanistic, terms, and attributed perception and consciousness to them. Her worldview, she said, "recognizes the perceptive capacity of all live beings."[6] She also said that "consciousness is a property of all living cells," even the most elementary ones: "Bacteria are conscious. These bacterial beings have been around since the origin of life."[7]

With allusion to the way in which this perspective can dissolve the mind-body problem created by mechanistic views of life, Margulis said: "Thought and behavior in people are rendered far less mysterious when we realize that choice and sensitivity are already exquisitely developed in the microbial cells that became our ancestors."[8] If carried through to its logical conclusion, this line of thought would have led Margulis to avoid the mind-body problem by joining us in affirming the position that we call "panexperientialism."[9]

The Direction of Evolution

From our Whiteheadian perspective, organisms are not simply, as neo-Darwinists typically portray them, objects that are shaped by randomly mutating genes and then selected by the environment. Rather, we see the activity of organisms as capable of playing significant roles in determining the direction taken by evolution. Whereas neo-Darwinists generally presuppose a unidirectional causality, according to which the behavior of organisms is determined by their most elementary constituents (understood mechanistically) and the environment, we "expect causal efficacy to flow in all directions."[10]

This way of thinking is supported by Margulis's view that life at all levels is composed of choice-making organisms. She wrote: "Perception, choice, and sensation apply not just to human beings or animals but . . . to all life on Earth."[11] In support of the notion that "free will may . . . be nature-deep," she said: "Aware of shape and color," large single-celled forams "make choices and reproduce their kind."[12] Living systems, she said, "are autopoietic: they are self-forming, or at least self-maintaining."[13]

In fact, having learned something about process philosophy, Margulis explicitly endorsed its view that organisms, with their choices, play evolutionary roles. Speaking of a "point of interest . . . from a process philosophy perspective," she wrote: "Choices made by organisms—what to eat, where to live, and with which organisms to associate—seem to be able to translate directly into the genetic effects of evolution. . . . Darwin . . . ascribed choice to worms, to female insects and birds choosing mates, and so on. In sum, animal choices . . . by social association leading to symbiosis and potentially to symbiogenesis—would seem to have a real role in evolution."[14]

Compound Individuals

In developing his philosophy on the concept of organism, Whitehead affirmed that there are not only "basic organisms," which do not include more fundamental ones, but also "organisms of organisms." Electrons and hydrogen nuclei are quite elementary organisms; then "the atoms, and the molecules, are organisms of a higher type, which also represent a compact definite organic unity." Still more complex are "individual living beings."[15]

The term for this notion was provided by the second most important process philosopher, Charles Hartshorne, in a 1936 essay entitled "The Compound Individual."[16] The basic idea is that one individual, such as an atom, can be present in a more complex individual, such as a molecule, which can in turn be present in a still more complex individual, such as a cell, which can in turn be present in a still more complex individual, such as a squirrel or a human being. One of the great errors of traditional philosophy, Hartshorne said, was to assume that all individuals had to exemplify the idea of an "individual" in the etymological sense of the term—that is, "indivisible," "without parts"—so that there could be no compound or composite individuals.[17]

Coming to the realization that "live small cells reside inside the larger cells,"[18] Margulis worked out a similar notion as a scientist, saying: "Different bacteria form consortia that, under ecological pressures, associate and undergo metabolic and genetic change such that their tightly integrated communities result in individuality at a more complex level of organization." In such developments, the joining together of individuals at one level can create "a new whole that was, in effect, far greater than the sum of its parts."[19]

Margulis thereby came to the idea for which she is now most celebrated: that (in Richard Dawkins's words) "the eukaryotic cell is a symbiotic union of primitive prokaryotic cells."[20] Eukaryotic cells, in other words, are composite individuals. The title of an essay describing the "transition from bacterial to eukaryotic genomes" refers, in fact, to "composite individuality."[21]

The basic question that needed to be answered, she said, was: "How . . . do independent, separate organisms fuse to form new individuals?"[22] Expressing the importance of this concept for biology, she wrote: "Failure to acknowledge the composite nature of the organisms studied invalidates entire 'fields' of study."[23]

We wonder whether, if the scientific world had accepted Whitehead's proffered framework for science, Margulis's idea of the eukaryotic cell would have faced the widespread ridicule with which it was greeted when it was first proposed.

Lynn with her father,
Morris Alexander, in Chicago

With her little sister and
mother in Chicago

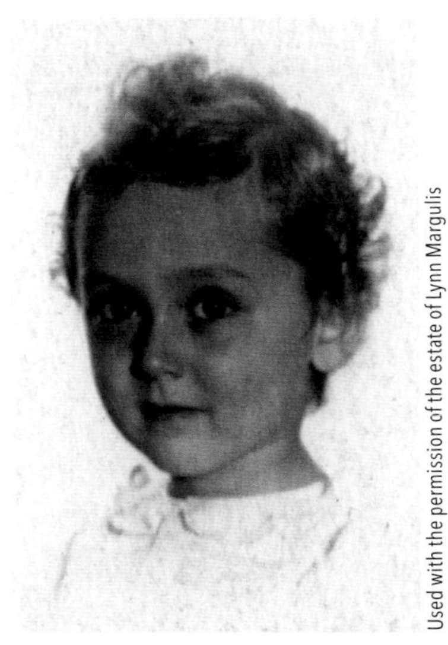

A formal baby portrait of Lynn
in 1939 or ’40, not yet two

A young Lynn Alexander

Lynn with mother and sisters at her first wedding.
L to R: Dianne, Leone, Joan, Lynn, and Sharon

Relaxing with Carl Sagan, Peter Pesch, and Kameshwar and Kashi Wali on the University of Chicago campus

Used with the permission of the estate of Lynn Margulis

With Dorion in Madison, Wisconsin

Elsa Dorfman

Portrait by professional photographer and swimming friend Elsa Dorfman

Used with the permission of the estate of Lynn Margulis

Lynn with Dorion and Jeremy in Watertown, Massachusetts, in the early 1960s

By the volkswagon bus ready for a family trip. L to R: Jeremy Sagan, Jennifer Margulis, Lynn, Zachary Margulis, Dorion Sagan, and Nick Margulis

On the phone at Boston University c. 1980

BU Archives

With Ricardo Guerrero at Boston University

Used with the permission of the estate of Lynn Margulis

Betsey Dexter Dyer looking through Dorion and Lynn's *Garden of Microbial Delights*, 1988

BU Archives

Lynn with James Lovelock

In the 1970s with James Lovelock in front of his Gaia statue in his garden in Cornwall

Courtesy William J. Clinton Presidential Library

Bill Clinton awarding Lynn the National Medal of Science

James MacAllister

Sharing a moment with Sean Faulkner

James MacAllister

Holding court with students at Harvard Forest in Petersham, Massachusetts

James MacAllister

At Puffers Pond in
Amherst, Massachusetts

Leonardo da Vinci Society

In Tempe, Arizona,
as the 2010 inductee into
the Leonardo da Vinci Society
for the Study of Thinking

Sean Faulkner

Working with Mohammed Et-Touhami in Morocco.
This "Rosetta" rock has three distinct geological deposits.

Gaia and Downward Causation

Given Whiteheadian process philosophy, downward causation from biological to physical entities is not forbidden. Indeed, it is expected: there is reciprocal causation at every level of nature.

Many people who have celebrated Margulis's explanation of eukaryotic cells, and even her more general doctrine of symbiogenesis, draw back from her endorsement of Gaia, according to which the planet's life as a whole regulates some of its physical processes.

Part of the explanation for this rejection of the Gaia perspective seems to be the fact that although she and Lovelock did not regard Gaia as "an Earth goddess," and even though she did not say that "Earth is an organism" (contrary to Lovelock), the "Gaia hypothesis is a biological idea."[24] The idea that the distinctively biological level of nature could influence the physical levels conflicts with the scientific community's leaning toward reductionism, according to which biological organisms are always caused by—never causative of—physical entities. One critic, for example, "denies environmental homeorrhesis by the biota," saying that "only physical and chemical processes regulate temperature, acidity/alkalinity, and chemical composition of reactive gases of the lower atmosphere."[25]

By contrast, we as Whiteheadians say that although the Gaia notion could not have been predicted, it also cannot be ruled out and is, in fact, consonant with Whiteheadian ideas.

Competition and Cooperation

Now, having considered Margulis's views of organism, the direction of evolution, compound individuals, Gaia, and downward causation, we move to a central question with regard to spirituality: Does evolution entail that the hope for a peaceable civilization is necessarily futile?

Discussing the "riddle of the universe," Whitehead rejected the view that evolution shows cooperation rather than competition to be the route to success. But he also rejected the opposite view, according to which evolution supports social Darwinism. Rather, Whitehead said, the universe has "its aspects of struggle and of friendly help," so that "romantic ruthlessness is no nearer to real politics, than is romantic self-abnegation."[26] Because biological evolution is neutral on this issue,

Whitehead held the hope that human societies might move, as the title of one of his chapters put it, "From Force to Persuasion."[27]

Margulis expressed a similar outlook, saying that "humanity's spiritual and moral qualities" do not contradict biological evolution. She criticized what she called "the neo-Darwinist overemphasis on competition among selfish individuals—who supposedly perpetuate their genes as if they were robots." Against "the neo-Darwinist zoologists who assert that the accumulation of random genetic mutations is the major source of evolutionary novelty," she said, "More important is symbiogenesis, the evolution of new species from the coming together of members of different species." Indeed, she added, "the only documented cases of the 'origin of species' in real time involve not selfish genes but 'selfless' mergers of different forms."[28]

Speaking directly to the issue of the lesson of evolution for war and peace, she and Dorion Sagan said: "Life did not take over the globe by combat, but by networking."[29] In the new (1996) preface to their 1986 book, they wrote that "although we would be foolish to propose that competitive power struggles for limited space and resources play no role in evolution," nevertheless: "*Microcosmos*, in contrast to the usual view of neo-Darwinian evolution as an unmitigated conflict in which only the strong survive, more than ever [because of recent discoveries] encourages exploration of an essential alternative: a symbiotic, interactive view of the history of life on Earth." They added that "it is folly not to extend the lessons of evolution and ecology to the human and political realm. Life is not merely a murderous game . . . , but it is also a symbiotic, cooperative venture in which partners triumph."[30]

Critics have often used caricatures in order to criticize Margulis. Daniel Dennett portrayed Margulis as saying that "cooperation is the norm" so that "nature is fundamentally cooperation." George Williams said that she saw nature as "benign and benevolent."[31]

As we read Margulis, however, she did not make such statements. Rather, given the prevalence of the "nature-red-in-tooth-and-claw" picture of evolution, she tried to give a more balanced view. She did not say that evolution is *only* or even *fundamentally* cooperative but that it involves cooperation *as well as* competition. She did not deny that "life is a murderous game" but said only that "life is *not merely* a murderous game." As we read her, therefore, she gave the same kind of balance that Whitehead did.

Courage

Both Lynn Margulis and Alfred North Whitehead have stood out not only because of their brilliance but also because of their courage—the courage to challenge ideas about which there was virtual unanimity among their peers. Professor James Crow, with whom Margulis had studied genetics at the University of Wisconsin, recently said of her that besides seeing something new, "maybe more important than just seeing, she was quite willing to say so and to report on it."[32]

Besides showing this courage with regard to evolution, she also demonstrated it in relation to other issues. One scientific-political issue with which we have been involved is 9/11.[33] After attending our 2004 conference on evolution, she on the flight home began reading a book on 9/11 by one of us.[34] After further study in her spare time (!), she wrote in 2007 that 9/11 was a "false-flag operation, which has been used to justify the wars in Afghanistan and Iraq as well as unprecedented assaults on . . . civil liberties."[35] In a superb 2010 essay on the destruction of the World Trade Center, she spoke of "overwhelming evidence that the three buildings collapsed by controlled demolition," so that "the petroleum fires related to the aircraft crashes were irrelevant (except perhaps as a cover story)."[36]

Conclusion

Spirituality has to do with the human "spirit." A deep spirituality aims at truth, even when others do not wish to hear it. It is not silenced by ridicule or even by more critical defenses of the majority position. Those of us who follow Whitehead know what it is to adhere to unpopular ideas simply because they seem to us to be true. In Lynn Margulis, we found not only a fellow spirit but one of unusual depth. We celebrate her memory and seek to emulate her spirituality.

David Ray Griffin, professor emeritus of Claremont School of Theology and Claremont Graduate University, and a director of the Center for Process Studies, is one of America's foremost Whiteheadian philosophers and the author of multiple best-selling books focusing public attention on the scientific inadequacy of the government's account of the events of 9/11.

John B. Cobb Jr., professor emeritus of Claremont School of Theology and Claremont Graduate University, and one of the directors of the Center for Process Studies, is one of America's foremost Whiteheadian philosophers and the author of many important books, including For the Common Good: Redirecting the Economy toward the Community, the Environment, and a Sustainable Future.

A Ferocious Intelligence

DAVID ABRAM

Iconoclastic, vivacious, intuitive, gregarious, insatiably and omnivorously curious, partisan, bighearted, fiercely protective of her friends and family, mischievous, a passionate advocate for the small, the overlooked, the taken for granted.

I first encountered the name "Lynn Margulis" as one of the two authors of a riveting paper in an issue of *The Ecologist* that I chanced to pluck from the magazine rack of a bookstore on Vancouver Island sometime in 1983. I couldn't put the thing down; I stood in the aisle of that store for a long time, reading their paper in deepening wonderment as various patrons tried to squeeze past or gave up and went around the other way.

I had recently returned to North America from a year spent wandering as an itinerant sleight-of-hand magician through Southeast Asia, living among and learning from traditional, indigenous magic practitioners in rural Indonesia and Nepal. I was reeling from the culture shock of my sudden return to the modern world from the breathing cosmos inhabited by those traditional magicians, for whom every aspect of the natural surroundings was alive, awake, and aware. Torn out of that animate terrain and thrown back into a world wherein each earthly presence seemed to be neatly defined, measured, and explained into oblivion, I was feeling hopelessly isolated and bereft here in my own civilization.

Poring over the pages while blocking the aisle in that bookstore, however, I learned with fascination of a scientific hypothesis suggesting that our planet was in some important sense alive—that the encompassing biosphere functions less like a clutch of mechanically determined processes than as a coherent, spherical metabolism. From those pages I gleaned that there were researchers in the natural sciences whose discoveries were finally encroaching upon assumptions shared by the various *dukuns* (medicine persons) I'd hung out with in Indonesia and the several *dzankris* I'd lived with in Nepal. For the first time since leaving the so-called undeveloped world, I began to relax.

The other coauthor of the article was one "Dorion Sagan"; reading his byline beneath the text I discovered that Margulis's collaborator was also a sleight-of-hand magician. Aha! Here was a potential way for me into this charmed circle of researchers. That evening I dialed "directory information" and asked for the phone number of a Dorion Sagan in the Boston area. Since "Dorion" is a pretty weird name, the number was easy to locate. A vaguely sardonic voice answered the phone. I mentioned the article and about being an accomplished sleight-of-hand magician myself.

Soon enough I was visiting Dorion at his home, trading sleight-of-hand secrets with this curious rogue who had the driest sense of humor north of the Mojave Desert. The day after I arrived, Dorion took me over to meet Lynn at her Boston U. laboratory, and the three of us went out for a meal. We became friends, which in the Margulian sense meant something at once easy and remarkably familial. Or at least that's how it felt — although I lived a couple states away from these folks and didn't get to see them all that often.

Pretty quickly Lynn put me in touch with James Lovelock—by simply handing him the phone, unexpectedly, when I was on the line with her. James and I soon met up when we both held forth at the first public conference on the Gaia hypothesis (entitled "Is the Earth a Living Organism?") at the University of Massachusetts in 1985. That year, drawing upon a fascination with perception that'd been sparked in me by my craft as a conjurer, I published a long essay on "The Perceptual Implications of Gaia" in *The Ecologist*.

> Perception, we must realize, is more an attribute of the biosphere than the possession of any single species within

> it. The strange, echo-locating sensory systems of bats and of whales, the subtle heat sensors of snakes, the electroreception of certain fish and the magnetic field sensitivity of migratory birds are not random alternatives to our own range of senses; rather they are necessary adjuncts of our own sensitivity, born in response to variant aspects of a single interdependent whole.[1]

Before long I found myself lecturing in tandem both with Lynn and with Lovelock, first in Britain, at the initial Gaia conference sponsored by *The Ecologist*, then at a large gathering of the American Geophysical Union, which had chosen the Gaia hypothesis as their conference theme. At each of these, I explored different philosophical implications of the hypothesis.

Visiting with Lynn in those early years, on one evening I mentioned my desire to do graduate research in cognitive ethology (the newly coined term for the study of animal cognition, pioneered long before by Jakob von Uexküll, Niko Tinbergen, and others). Instantly Lynn called up a colleague, Jelle Atema, and bundled me off to meet him at the marine laboratory at Woods Hole, where he was director and where he was researching the sensory biology of lobsters. I didn't end up working with Jelle, but it was an engaging encounter (both with him and with the crustaceans). Lynn introduced me to many other brilliant creatures, some of them straitlaced, like the climatologist Stephen Schneider, and others fairly flamboyant, like the biologist-philosopher Francisco Varela, and several of these contacts stuck fast and sank their claws into my own thinking. When she was invited to present a public lecture in the Boston University series "On Religion," Lynn proposed that I give the lecture instead, with herself as the formal respondent to my remarks. It was a nifty honor for someone just barely embarked, as I was, on graduate studies in philosophy and ecology down in New York. She went all out for those whose character and insights she believed in. Lynn was a not merely a theorist of symbiosis; she was a keen practitioner of it, forming close-bound and bonded alliances with every kindred soul she bumped into.

Among the sphere of scientists and theorists who were taking Gaia seriously in those days, there was a certain puzzlement regarding how to reconcile

Gaia with Darwinian theory. (Several biologists had thrown down the gauntlet a few years before, arguing that since there was no population of living planets, or Gaias, competing with one another for limited resources, Gaia could hardly have emerged through a process of natural selection. Since Gaia can't reproduce itself, it cannot properly be considered a living entity, at least not within a Darwinian frame of reference.)

I knew Lynn well enough at that time to call her in the middle of the night when I awoke from a dream with a simple recognition: "Gaia," I blurted to Lynn's obviously groggy voice on the phone, "is not something that undergoes natural selection. Rather, Gaia is the one who does the selecting. I mean, she—or it—is the phantom presence tacitly implied by Darwin's active metaphor of 'selection,' the self-organizing entity that does the selecting!" I thought it was a pretty cool insight, but Lynn just said "Okay."

"Huh?"

"Okay; now go back to sleep."

Another criticism of the hypothesis came from Stephen Jay Gould, whom I greatly admired. When I asked Steve about Gaia after a large lecture he gave at SUNY Stony Brook, he blustered: "The Gaia hypothesis says nothing new—it offers no new mechanisms. It just changes the metaphor. But metaphor is not mechanism!"

This seemed kinda goofy to me, since "mechanism" itself is nothing other than a metaphor. Steve was just affirming his allegiance to mechanical metaphors rather than organic metaphors. Many folks were looking to Gaia theory in hopes that it would generate a host of new mechanisms. But it appeared to me that Gaia offered something much wilder; that by shifting our understanding of the encompassing earth away from metaphors grounded in machines—that is, metaphors borrowed from things that humans design and build, and hence that we can readily take apart and put back together—toward organic metaphors rooted in bodies that grow and self-organize and improvise their way in the world, Gaia theory was opening a whole new approach for the natural sciences. I began to puzzle out the curious influence of metaphor on the practice of science.

> The scientist who holds a fundamentally mechanical view of the natural world must suspend his or her sensory participation with things. He strives to picture the world

> from the viewpoint of an external spectator. He conceives of the earth as a system of objective relations laid out before his gaze, but he does not include the gaze, his own seeing, within the system. Denying his sensory involvement in that which he seeks to understand, he is left with a purely mental relation to what is only an abstract image.
>
> Likewise with any particular object or organism that the mechanist studies. There as well, she must assume the position of a disinterested onlooker. She must suppress all personal involvement in the object; any trace of subjectivity must be purged from her account. But this is an impossible ideal, for there is always some interest or circumstance that leads us to study one phenomenon rather than another, and this necessarily conditions what we look for, and what we discover. We are always in, and of, the world that we seek to describe from outside. We can deny, but we cannot escape being involved in whatever we perceive. Hence, we may claim that the sensible world is ultimately inert or inanimate, but we can never wholly experience it as such. The most that we can do is attempt to *render* the sensible world inanimate, either by killing that which we study, or by deadening our sensory experience. Thus our denial of participation ultimately manifests as a particular form of participation, but one that does violence to our bodies and to the earth. . . .
>
> Mechanism sublimates our carnal relationship with the earth into a strictly mental relation, not to the world, but to the abstract image of a finished blueprint, the abstract ideal of a finished truth.[2]

That was hardly Lynn's style of science. She was passionately engaged with the myriad organisms she studied, participant in a living interchange between her own creaturely self and the multiple forms of sensitivity and wild sentience whom she hosted in her laboratory and whom she pondered, posed questions to, negotiated with.

> If the mechanical model of the world entails a mentalistic epistemology . . . the Gaian understanding of the world

> (that which speaks of the encompassing earth not as a machine but as a self-organizing, living physiology) entails an embodied, participatory epistemology. As the earth is no longer viewed as a machine, so the human body is no longer a mechanical object housing an immaterial mind, but is rather a sensitive, expressive, thinking physiology, a microcosm of the autopoietic earth. It is henceforth not as a detached mind, but as a thoughtful body that I can come to know the world, participating in its processes, feeling my life resonate with its life, becoming more a part of the world. Knowledge, ecologically considered, is always, in this sense, carnal knowledge—a wisdom born of the body's own attunement to that which it studies, and to the earth. . . .
>
> We may wonder what science would come to look like if such an epistemology were to take hold and spread throughout the human community. It is likely, I believe, that scientists would soon lose interest in the pursuit of a finished blueprint of nature, in favor of discovering ways to better the relationship between humankind and the rest of the biosphere, and ways to rectify current problems caused by the neglect of that relationship. . . . Experimentation might come to be recognized, once again, as a discipline, or art, of communication between the scientist and that which he or she studies.[3]

After a few years I stepped away from Gaia theory, disturbed by the facile manner in which it was being taken up within the culture at large. This had partly to do with its ungrounded appropriation within New Age circles. But also in more intellectual spheres, where Gaia was increasingly associated with global thinking, and globalism, at a time when the US government was strenuously pushing an agenda of economic globalization and free trade one-worldism. It was an era when much of the educated populace was uncritically lining up behind what seemed, to a few of us, merely a new form of market imperialism—a new license for the homogenization of human culture and the corporate plundering of every last part of this breathing planet. In contrast to such global thinking, with all the generalities it entailed, my attention was increasingly drawn toward local, place-based particulars and bioregional initiatives. Much of my work since that time has been dedicated to a rediversification of culture—not through any reversion to ethnic or religious traditions, but rather through the face-to-face participation of citizens in the replenishment of the local earth (a deepening reciprocity between human communities and the living landscapes that surround and sustain them).

Lynn's work suffered from no such need. Her contributions to Gaia theory, with that theory's large-scale systemic character and the overwhelming hugeness of its object (or subject), had always been balanced by her attention to the variegated specificity and inexhaustible diversity of what she and Dorion called the microcosm: the realm of the small ones who writhe and dance and feed within our tissues, carrying on their alchemic exchanges and negotiations within our intestines and our eyelids, in the gums of our mouths and across the mucosal secretions of our genitalia, as well as in the guts of the termites who chomp their way through the walls of our home, and teeming in the dank soil and the stagnant puddles along the sidewalk. Lynn was a partisan of ooze, a champion of the stuff that most people would prefer to ignore: here, too, she insisted, was life, sensitivity, sentience.

In this sense Lynn was an intermediary between human culture and the more-than-human community that thoroughly permeates and yet vastly exceeds the human world. Mischievous to the max, she took great pleasure in deflating human-centered arrogance and pretension, dislodging any comfortable sense of human uniqueness and transcendence. It was life—profligate, teeming *life* in all its weirdness—that held the magic for her, not this featherless biped with its confused aspirations. Lynn intuited and doggedly gathered evidence to show that most anything we two-leggeds take special pride in—our capacities for cogitation, conviviality, and culture—had been invented, eons before, by the microbial entities that compose us. Like other great biologist-intermediaries (Barbara McClintock, for example, or Jane Goodall), Margulis, with her odd and unexpected angle on the world, opened, and continues to open, an outrageously fresh perspective on evolution and ecology, a gestalt shift wherein matters conventionally hidden behind the agonistic struggle for survival take on new primacy and importance: reciprocity, cooperation, and mutual exchange.

Yet this was no mere reversal of priority from masculine to feminine modes, or from aggressive to receptive forms of relation. From Lynn's perspective, cooperation was precisely *a competitive strategy*, a canny and creative way to outcompete less sociable others. Receptivity (both in Lynn's work and in her person) was here combined with a kind of healthy, exuberant aggressivity, a lust or gusto for life. After all, many of her organisms enacted their coalitions by engulfing or consuming others, and so incorporating those others within their flesh. Gusto indeed, and please pass the salt!

Margulis seemed to have as little patience for current or ostensibly enlightened notions of the feminine and feminism as she had for stereotypes of masculinity. I've said that she was a practitioner of symbiosis, forming rich alliances and bonds with those around her. Yet she was never afraid to piss people off and rarely shied away from the chance to do so. I remember her scornfully stalking out in the middle of a panel discussion in which she was taking part, at a Lindisfarne Fellows meeting in New York City, after farmer-authors Wendell Berry and Wes Jackson mounted a moral critique of biotechnology (this was back in the years when large-scale genetic engineering was just getting under way). Yet I know that she soon came to have solid friendships with both those gentlemen.

One thing that sticks with me about Lynn: for many years she was probably the speediest thinker and the fastest talker I'd ever encountered: I had the sense that I had to be taking methamphetamines if I really wanted to keep pace with her zinging mind. That she began to slow down, somewhat, in her last years was probably a blessing to many of her friends, who maybe found that they could now relax and remember to breathe when they were talking with her.

Lynn was a wonderfully free thinker, often weighing in hard against whatever stance was conventionally assumed to be politically correct. At least in private, she had a withering disdain for neo-Darwinian dogmas. I remember her advocacy for the work of the early twentieth-century science-theorist Ludwik Fleck (and her sense that Thomas Kuhn had cribbed some of his main ideas from Fleck without at first giving much credit). I remember her admiration for Fidel Castro (how impressed she was with the near-universal literacy she found in Cuba when she visited there). I remember her personal disappointment upon finally reading

Peter Kropotkin's 1902 work, *Mutual Aid: A Factor of Evolution*. The great anarchist's tome had anticipated some of her own research and insights, yet although *Mutual Aid* was ostensibly a work on evolution, she felt that there was precious little real biology in the book. Although she was clearly a woman of the Left, for Lynn the dazzle of detail in painstaking biological observation always trumped politics.

We shared a conviction that sentience had nothing to do with an amorphous mind or spirit somehow lodged within a body—indeed that intelligence was not even an attribute of the brain but was rather the cognitive acuity of a whole body actively engaged with its world. Not just the human body, then, but the body of a swerving hawk or a paramecium or a corkscrewing spirochete. Glancing over my own writings in the last couple decades, I see very few that were not subtly influenced by Margulis and by the charmed sphere of poet-scientists and philosopher-researchers within which she circulated.

> I learned the complex chemistry of photosynthesis back in high-school, and studied it more intently in college, suitably impressed by the elegant efficiency of the process. But still I wonder: what does it *feel* like to be so rooted in a place, sipping minerals through root filaments that extend themselves, by taste, through the dark density underground, while drinking the sunlight all day with your needles? What is the precise *sensation* of transmuting sunlight into matter? Surely we don't believe that such a metamorphosis happens without any concomitant sensation—that there is no experience whatsoever accompanying this transformation!
>
> It seems obvious that both leafing and needled trees lack a central nervous system, and hence are far less centralized than we in their experiencing. Nonetheless, the fact that sensations are not referred to a central experiencer does not negate the likelihood that sensations *are* being felt in the leaves themselves. Experience, for most plants, is simply a much more distributed and democratic affair than it is for more hierarchically organized entities like us.
>
> We may wish to assume that a cottonwood tree is utterly void of sensation, and hence that a summer sunrise makes

no impression within its flesh. We may convince ourselves that there is no feeling within its leaves that might distinguish an afternoon of steady overcast from an afternoon without clouds, no sensations at the tips of the rootlets when a drenching rain penetrates the soil, or within the woody sheath of the cambium as the sap rises within the trunk, and that indeed all the daily and seasonal shifts within a tree's metabolism unfold according to a purely mechanical causality, without any *need* of sensation and indeed in a complete absence of impressions—in a blank, vacant expanse of mute materiality.

But such a notion implies that our own capacity for experience is a sudden arrival in the material field. It ensures that our own sentience cannot be thought of as an elaboration of a sensitivity already inherent in organic matter—cannot be felt as the unfurling of a responsiveness already present, for instance, in the myriad microbial entities whose collective activities enable much of the vitality of plants and animals. It implies, rather, that awareness is a power that abruptly breaks into bodily reality from elsewhere.

If, however, I allow that this wild-fluctuating sensibility I call "me" is born of this upright body as it improvises its way through the world—if I acknowledge that my sentience is supported by the air streaming in through my nostrils, as well as by the manifold sensibilities that move within me (by the keen responsiveness of the bacteria in my gut, and the skittishness of each bundled neuron within my spine)—then a new affinity with the sensuous world begins to blossom. For now the other bodies that I see around me—whether blackbirds or blades of grass, or the iridescent beetle currently crawling across my shirt—all give evidence of their own specific sentience.

The emerald leaves dangling like butterflies from the near branch of an aspen attest by their very hue to a kind of ongoing enjoyment along the fluttering periphery of the tree—an exaltation of chlorophyll. As though one's breathing lungs were flattened and spread out across the smooth surface of

> one's skin, and the day's warmth brought a tingling transmutation along that surface, one's outermost membranes being ravished by the rays, from dawn until dusk.
>
> Looking up, I notice the needled hillside across the valley now as a curving field of sensations—since my skin, at this moment, feels the variegated green of all those trees as a quiet ecstasy riding the hill. It is an ecstasy I regularly partake of by receiving the radiance of that color within my eyes, a gentle edge of pleasure that has always been there for me in the green hue of leaves and of clustered needles—a subtle delectation in the sight of green, felt much more intensely whenever sunlight spills across the grasses or the leafing trees—but which I've become fully conscious of only now, as *a kind of empathy in the eyes.*[4]

When Lynn and Dorion stayed with us in New Mexico a few years ago (during a meeting of the newly revived Lindisfarne Fellowship), I remember the abundant warmth and attention that Lynn paid to my partner, her interest in my young children, and her keen curiosity about whatever was unfolding in my life. She was a bighearted soul, always able to shift easily between the microcosm and the macrocosm—from the proliferant wildness of the microbial world to the vast and threatened integrity of the biosphere—and to return instantly from either back to the intimate empathy of human friendship. Truth is, Lynn could participate in each of those divergent scales at the same time—which everyone does, I suppose, but she did it consciously, and with an intensity that inevitably altered, and left her signature trace within, all those dimensions.

David Abram, cultural ecologist and geophilosopher, is the author of Becoming Animal: An Earthly Cosmology *and* The Spell of the Sensuous: Perception and Language in a More-Than-Human World.

Fishermen in the Maelstrom:

Big History, Symbiosis, and Lynn Margulis as a Modern-Day Copernicus

PETER WESTBROEK

When Lynn was in her early forties, she confided to me that her end was nearby. Once her message was delivered, she would simply slip away. How lucky are we then, considering the sheer magnitude of this message, that she would be with us for another thirty-odd years. And despite the importance and multifaceted nature of her work, it is ironic that not even a one-liner need sum it up. Rather, a single word suffices—symbiosis. For Lynn, symbiosis was not a biological curiosity, as the majority of her colleagues would have it, but the very heart of life.

Lynn's radical perception of symbiosis proved to be a bone of contention. We know how her early manuscripts were rejected time and again, and what desperate fights she had to wage for recognition. We should not be surprised that she deeply loathed the arrogant self-glorification of the scientific establishment. This judgment would stay with her until the end of her days, even if by then she was overloaded herself with the most prestigious honors and awards. For example, as late as April 2011, I attended an honorary public lecture by Lynn for the centenary of the Netherlands Society of Microbiology. The Prince of Orange was there, as well as an exquisite collection of Nobel laureates and other distinguished personalities. Afterward, my friends inquired why her talk had been so defensive. After all, she had received this invitation as a token of high appreciation in the country where microbiology had been invented with Antonie van Leeuwenhoek, the Dutch tradesman from Delft who handcrafted the first microscopes and was the first person to see protoctists, sperm cells, and bacteria.

How and why did Lynn's crusade for symbiosis turn out to be so remarkably successful, more so than she could have initially imagined? Some people would say that her success was due to her unrelenting fighting spirit. Others may point to feministic overtones of the symbiotic concept. She herself might have said that, like all decent scientists, she stood on the shoulders of giants, merely extending the earlier work of predecessors like Darwin, Wallin, Merezhkovsky, and Famintsyn, and that in addition she could rely on her immense network, her students and colleagues, and above all her children and grandchildren. All of this is true, of course, but I believe something even greater is in play.

I see Lynn as the herald of a new worldview, a major landslide in public orientation that swept through the world at large during her lifetime. This change is at the same level as the transition from the geocentric to the heliocentric worldviews, which brought us modern science. But in this case, it is the symbiotic worldview.

Involvement and Detachment

In his seminal book *Involvement and Detachment*, sociologist Norbert Elias[1] regards all major worldviews as revolving about the relationship between humans and their environment, in particular the earth. He uses a story by Edgar Allan Poe, "A Descent into the Maelström," to elucidate his point. Three fishermen are caught with their ship in a huge and devastating whirlpool and dragged into an ominous spiral course toward the abyss, round and round, almost horizontally, along the sloping wall of water. One brother is pulled down and disappears. The other brothers know that they are doomed, but how different are their reactions! One closes his eyes and waits for the end, while a flash of remarkable clearheadedness seizes the other. Fascinated, he observes the numerous objects drifting in the water around him, most of them dilapidated remains of shipwrecks. When he sees that elongate, cylindrical objects remain unscathed, he discerns a small window of opportunity for survival. He warns his brother, but this poor man is unable to respond. This leaves the clearheaded fisherman with no other choice than to hang on to the empty water cask to which he is attached, cut it loose, and throw himself with it into the boisterous waters. Below him, he can see the ship with his beloved brother diving down into the abyss. Supported by the water cask, he continues to swing around in the whirlpool until the hurricane dies down. The water comes to rest, and the fisherman can escape.

Elias sees the dynamics of worldviews as resulting from a subtle interplay between involvement and detachment. People are "involved" when, overwhelmed by their fears, they are left to the dynamics of their surroundings, unable to act in any purposeful way. They are disoriented, escape into emotions, and give their fantasies free rein. This attitude makes them all the more vulnerable. The fisherman who loses his life in the maelstrom exemplifies this state of involvement. The "detachment" of the survivor exemplifies the reverse situation: people with minds open to reality can acquire an understanding that allows them to adjust challenging situations to their advantage. Detachment requires the suppression of fear, the acquisition of orientation, and a balanced state of mind. The involvement does not disappear but is curbed to become the driving force of purposeful action.

Punctuated Evolution of Worldviews

As societal phenomena, worldviews are more encompassing than isolated scientific breakthroughs or currents of intellectual speculation. The reason why involvement and detachment play an important role in their dynamics is that they are rooted in the power structure of the societies to which they belong. As long as they agree with or even support the prevailing societal structure, they form islands of detachment amid an ocean of chaos and involvement. However, as people accumulate more experiences with their environment, their ideas will eventually exceed the accommodation space of the prevailing worldview, so that a conflict with the associated power structure will become apparent. It does not come as a surprise, then, that the history of worldviews is punctuated with wars and outbursts of involvement. Subsequently, a new association of a worldview and a power structure will emerge, providing a new island of detachment.

The advent of the heliocentric worldview may serve as an example. It chased humankind out of the comfortable position it had occupied under the preceding geocentric regime and placed us in a marginal position, together with our newly marginalized planet. For the first time, nature emerged in the conscious mind as an independent, indifferent world.

The passage toward the heliocentric worldview was not instantaneous. Its unfolding took centuries. While rudiments of the geocentric worldview persist until the present day, science and technology carried the marginalization of humans further ahead. The history is well known: the earth, previously the holy theater of God's history with man, was mapped out and exploited; the anatomy and physiology of the human body were systematically investigated. Later on, the first geologists explored the fathomless depths of time. Darwin's evolutionary theory reduced humanity to an animal species. Sigmund Freud demonstrated that our emotional world is the scene of unregulated and unconscious drives. And modern astronomy chased even the sun out of its central position. Heliocentrism, once the point of departure for this development, steadily changed into the *modernist* vision of reality from which all vestiges of a center had vanished.

But even as heliocentrism took root in the human psyche, one stronghold persisted. Humans remained the masters of the earth.

Birth of the Symbiotic Worldview

Back in December 1968, the *Apollo 8* astronauts, wrapped up in their robotics, shot into space. When they looked back, they saw a new planet, at once familiar and unknown. The famous *Earthrise* photograph from deep space[2] became the icon of a new element in our emotional lives, of a worldview endowed with respect and amazement for the earth's complexity.

The impact brought about by *Earthrise* would probably not have been as powerful if that photo had reached us at another moment. For twenty years, the Cold War had kept the world in its iron grip. The Cuban missile crisis made people realize that for all those years they lived at the brink of annihilation. The Vietnam War showed the moral bankruptcy of the West, while funds that could be used to dispel the hunger in the world were diverted to the production of weapons. There was a growing demand for a new perspective. If we were able to travel to the moon, there was no problem we couldn't solve.

But the space images also showed quite another, more painful reality. From the vantage point of deep space, humanity became an almost invisible yet potentially ravaging monster, a threat to this unique oasis in space. Soon after *Earthrise,* the exhaustion of natural resources, climate change, overpopulation, and the destruction of biodiversity moved to the top of political agendas.

Although hell has not yet broken loose, this uninterrupted bombardment of depressing predictions has a dangerously disorienting effect. A nondescript but acute global anxiety is mounting. The epidemic of involvement, as in Poe's story (without the needed scientific palliative of a long view of the real world) is inside ourselves and forms the greatest danger of all, with outbreaks of outright denial, indifference, intolerance, fundamentalism, nationalism, racism, and xenophobia.

Our present situation is an early phase in the transition from one period of relative stability into the next. New experiences with the world give rise to an emergent worldview. This worldview is in conflict with the old power structure still in control. Under the circumstances, scientific developments play an important role in exacerbating this contradiction.

Earth system science,[3] or Gaia science, as Lynn would have it, represents a profound metamorphosis in our perceptive capabilities, partly because it considers the full 45 million centuries of our planet's existence. Leaving temporarily aside the towering problems around, it makes us acquainted with the unique position of our planet in the solar system, unravels the incredibly long and dramatic history from which we ourselves have emerged, and investigates the position of us humans in this great planetary perspective.

Geology Before the Symbiotic Worldview

Geology, the science of the earth par excellence, is a suitable dipstick by which the emergence of the symbiotic worldview may be estimated. When I was a student, in the late 1950s and early '60s, geology ignored its own subject. The few geologists who speculated about global matters were dismissed as "geopoets," as there were insufficient data to prove them right or wrong. The hope was that by combining all the regional knowledge, the large planetary overview would eventually emerge, just as the globes had once been puzzled together from individual maps. We also knew very little of geological history. Most geologists focused on the portion of earth history with clearly recognizable fossils. As it turned out, this covered only the last bit of geological time—about one-eighth—and all the rest was largely ignored. It was left mainly aside because in the absence of identifiable fossils, the rocks were usually hard to put dates on. Also, we were ignorant about the oceans, although they covered two-thirds of the planetary surface. Furthermore, geology was highly descriptive and split up into a steadily growing multitude of poorly connected disciplines. Life was generally regarded as only adapting to the changing physical and chemical conditions on earth, and there was little consideration for its participation in global dynamics.

One wonders: was it any fun to be tied up in such a field? Let me assure you: it was fabulous. As we were in the middle of it all, we didn't notice any of these shortcomings. We were not bothered in the least with global dynamics or all of earth history. What we had on our hands were fascinating puzzles about the history of the mountains in Spain or the Alps, about the ice age in the Netherlands, or the fossils in the Devonian period. My appreciation of the state of geology in the late 1950s and early 1960s is no more than hindsight knowledge. The most ironic thing of all is that we didn't have the slightest notion of the giant revolution that was about to happen. It was like New York on 9/10, business as usual.

The Transition to Earth System Science

It was in 1967, one year before the *Earthrise* photo was taken, that the plate tectonic theory was born. In one stroke it swept most of the old fantasies on global dynamics into the dustbin. For the first time, geology focused on the dynamical history of the entire planet. Small-scale phenomena became integrated components of a global master scheme. Plate tectonics described the dynamics of the solid earth in physical terms. Geochemical models of earth dynamics soon followed suit. The earth's chemicals were thought to circulate through the maze of tubes between reservoirs. The element carbon, for example, is distributed between the trace amount of carbon dioxide in the atmosphere-ocean system and the huge lithospheric reservoirs of limestone and organic carbon. The geochemists were able to demonstrate an ongoing redistribution of the total carbon over these reservoirs on the multimillion-years time scale, causing these reservoirs to "breathe" in a coordinated fashion. One can imagine that writing geological history in those terms was a real sensation. These developments would not have been possible without the enormous advances in geohistory analysis, which allowed the entire geological record to be dated and interpreted with enhanced detail.

Finally, geobiology came to the fore as a major component of earth system science. Life was now considered not only to adapt to environmental changes, as it was in the past, but also to act as a major geological force, influencing in multiple ways the dynamics of the planet.[4] It is within this burgeoning field that the research of Lynn Margulis was rooted. To her, the global dimension of symbiosis was a matter of course, so that she could team up with James Lovelock without hesitation.[5] To her, the earth was the largest of all ecosystems, with the biota sharing and exchanging energy, water, and nutrients with its environment. Not only did she infuse the Gaia theory with her deep biological knowledge and intuition, but we also owe much of its remarkable success to her exceptional communicative gifts.

New Methodology

Under the modernist worldview, science had been dominated by the Cartesian, or analytical, method. From the dazzling wealth of phenomena around us, Descartes sought to isolate those elements that could be understood in the divine light of the ratio, of clear, logical thought. To achieve this goal, he had to definitively divorce subject and object. The complex reality where all things influence each other was unsuited for rational understanding. Only the objective investigation of reality could provide the understanding of the world that was logical and acceptable for all. Central to this method was the scientific experiment. A single element was separated from its environment and subjected to a single influencing force. The result of a successful experiment was a logical, causal, and reproducible relationship of general validity, an inalienable heritage for humanity. Although the success of this method has been, and still is, overwhelming, it atomized our understanding of the world.

It was soon after the *Earthrise* photograph was publicized that the Cartesian method of analysis was complemented with synthesis, in order for the complexity of the real world to be addressed. In his monumental *La méthode*,[6] the French philosopher Edgar Morin analyzes complexity in all its manifestations, from elementary particles to biological organization and human relationships. He argues that the real world is not complex just because it is complicated. Relationships are at the same time antagonistic and complementary. Cause and effect are opposed and at the same time intimately related. Life adapts to and changes the environment. The brain produces thoughts that in turn change the brain. Nature generates culture and culture changes nature. Disorder makes order and the other way round. It is the emphasis on this paradoxical nature of reality that changes our view: whereas the science of the modern worldview took things apart and studied the fragments in isolation, the science of the symbiotic worldview places them into their context.

These different approaches of analysis and synthesis are not mutually exclusive. Again, the relationship is complex—antagonistic *and* complementary. The analytical approach calls for synthesis, while synthesis would devolve if it were not kept in check by an adequate body of analytical research. Always the two go hand in hand.

Outcome and Implications of Earth System Science

Among the major topics in earth system science are long-term changes in tectonic regimes and continental masses, the global cycling of chemical elements, oxygenation of the atmosphere and the oceans, global climate dynamics, and changing patterns of global biodiversity. The origin of life and biological evolution are understood as emergent features of earth system dynamics. Progress depends on the spectacular increase in sophistication by which the geological record can be interpreted. The general approach is to quantify the results of this fieldwork and to integrate them into mathematical models. On the outside, the ambition of earth system science may seem far-fetched. But by now, fifty years after the emergence of plate tectonics, it has become a major focus within the natural sciences.

To epitomize the outcome of earth system science so far, the metaphor of a ratchet springs to the mind, an asymmetrical cogwheel-and-click that can turn only in one direction. Ratchets are used in clockworks, jacks, rattles, and mechanical toy trains. I use the term to denote irreversible, accumulative developments in complex systems. The ratchet is, of course, a poor metaphor. In contrast to this mechanical analogue, the earth builds up its complexity, differentiation, and information content throughout geological history. We witness an ongoing stream of new attributes emerging, against a countercurrent of information loss. Another difference is the unpredictability of the earth's development. For example, the planet is known to have passed through a number of deep crises, when it seemed to return to earlier stages of its development. Time and again, however, the system veered back with a surprising speed and subsequently regrouped, following a different course than before. On such occasions, it seems as if a global memory was in operation.

The "ratcheting" behavior has been documented for many phenomena, including biological evolution, mineral diversity (from a dozen minerals in the original dust cloud from which the solar system emerged to more than 4,300 in the present Earth), atmospheric oxygen levels (from 0 to 21 percent), as well as the complexity of global geochemical pathways and their regulation. Thus, overall the environmental diversification and differentiation appear to have been on the increase, from the origin of the earth up to the present day.

It is important to realize that early in earth history this differentiation must have been purely physical and chemical. The agglomeration of the early Earth as a separate component of the solar system; the segregation of the moon; the subsequent differentiation between the core, the mantle, the crust, the hydrosphere, and the atmosphere; the emergence of initial tectonic regimes, which in turn gave rise to a steadily proliferating spectrum of minerals and rock types—all these early differentiation steps took place in the absence of life, independent of Darwinian mechanisms. The first living systems likely emerged locally out of complexifying elaborations of geochemical fluxes. This event marked the origin of miniature, self-replicating organizations—the first cells—that would soon spread out over the entire planetary surface and increasingly modify previously existing mechanisms of earth dynamics. Life, indeed, represents an immense geological force,[7] and with the advent of biological evolution a threshold was overcome after which the rate of global differentiation highly increased. Thus, life cannot be considered as an autonomous process adapting to the physical and chemical constraints imposed by the earth system. Instead, biology is part and parcel of the ratcheting earth or, to put it more bluntly, an emergent property of the planet, as it speeds up the rate of its differentiation.

Cultural evolution (or the "civilization process" in Norbert Elias's terminology[8]) stands out as the subsequent major revolution in earth dynamics. In the last 200,000 years of earth history, complexity locally increased to the point where the ratcheting became more and more independent of tedious Darwinian mechanisms.[9] The new principle proliferated slowly at first, but then at a rapidly increasing rate, until presently it modifies the dynamics of the outer earth. Its differentiating effect on the planet (to be expressed in tools, thoughts, means of communication, and institutions) is immense.[10] Thus, from the point of view of earth system science, the civilization process is not the privilege of humanity but, like biology, an emergent property of earth dynamics. It is the earth that civilizes, through us. Consequently, the civilizing process cannot be divorced from the chemical, physiological, and biological factors governing earth dynamics. Like biological evolution, the civilizing process amplifies and speeds up the ratchet of the earth's differentiation.

According to Elias,[11] increased involvement is the main reason why the natural sciences progress so much faster than the human sciences. But there may also be a methodological difficulty. We have to become accustomed to the idea that even our most intimate feelings and relationships are ultimately geological phenomena. This implication of earth system science does not flow from excessive reductionism but demonstrates the extreme degree to which global differentiation has advanced. Thus, even our ego is not monolithic but paradoxical and complex *sensu* Morin. On the one hand, it secures the integrity of our personality, on which we depend for our survival and emotional lives, while at the same time it is delusive—yet another source of our perennial tendency toward anthropocentrism. It seems to me that the top-down systems approach of earth system science will eventually clarify this paradox, so that consilience between the human and natural sciences may emerge at long last.

Orientation

This leads to an important and controversial fruit of the unfolding symbiotic worldview—the growing influence of science in shaping our orientation in life. In the modernist age, when the analytical method was overwhelmingly dominant, science could hardly provide a coherent picture of reality. All it could do was to correct isolated notions drawn from traditional wisdom. No, the sun does not turn around the earth, but it is the other way around. No, the earth is not 6,000, but 4.6 billion years old. No, life and biodiversity did not result from divine creation but emerged from immanent causes.

By contrast, religions and unchecked myths provided a wealth of wonderfully coherent stories, deeply rooted in the souls of the millions. However, when the balance of science tipped over to the synthetic side, and the bits and pieces of science began to merge, realistic stories emerged that dwarf the wildest fantasies. Even though the results of science are replete with uncertainties, there is no way to get closer to reality.

Conflicting Worldviews

Already the symbiotic worldview is changing our perception of the world around. While in the 1950s we could still believe that the human race was destined to master this planet, earth system science reveals that we are entangled in a larger, autonomous process of global differentiation. Ultimately, we shall have no choice but to accommodate to the vicissitudes of this planetary odyssey, like surfers amid the ocean waves. Our humiliation appears to have reached its limits, and this in itself is enough reason for increased involvement. It lures us into building new strongholds of anthropocentrism. While we tend to protect ourselves from the savage world by spinning cocoons around our personal lives, our "civilized" society seems to opt for ignorance. Obsessed with personality, we indulge in our twittering global village. We are like goldfish swimming in a glass bowl, imagining ourselves to be in charge of the world ocean.

The marginalization of humanity was the bone of contention for the anthropocentrism of the geocentric worldview, because it put into question the foundations of the prevailing feudal system. Can we identify the bone of contention that brings the symbiotic worldview in flagrant conflict with the underpinnings of today's society? By itself, the shocking exposure by earth system science of our ego as an indispensable yet misleading delusion may still be dismissed as an intellectual sophistication. But the implications are less innocent. Whenever the same delusive ego seizes a portion of the system earth in order to subject it to his exclusive expedience, a mismatch with the symbiotic worldview seems in order. I conclude that the symbiotic worldview opposes the very heart of the present social order by exposing property, private or public, as plunder of the earth system. A new social order, in agreement with the symbiotic worldview, will have to cope with this inconvenient truth.

We must admit that leaders, nations, superpowers, and even the United Nations can never solve our problems. All these attributes are part and parcel of the fishbowl, the man-made cocoon in which we are locked up. Ultimately, no other refuge remains for our orientation than our civilizing planet itself. To keep the right course on our surfing voyage, we must trust the earth and use our science to sense its predispositions.

Margulis's Contribution and the Advantages of Postulating the Symbiotic Worldview

In hindsight, this essay may help us understand why Lynn Margulis's career made a difficult start. Her first publications appeared around the same time as the *Earthrise* photograph, when scientific analysis had the upper hand. The concept of symbiosis militated against the spirit of the time. Biological objects were usually studied in isolation, while Lynn placed them back into nature. An anastomosing network replaced Haeckel's phylogeny of branching lineages. Following the Canadian biologists Sonea and Panisset,[12] Lynn and her son Dorion Sagan regarded all the prokaryotes together as one single species. Prokaryotes, snot, and slime were and always had been the backbone of the biosphere—a global wrestling and copulating garden of microbial delights. They did not recognize in animals and plants a necessary higher status than members of the bacterial world, as all organisms had undergone evolution for the same length of time. Neither did they dismiss Williamson's theory of animal metamorphosis through fusion of unrelated organisms.[13] In their view, humans also became hybrids with a questionable identity—yet another blow to anthropocentrism. Even if by now there is no general acceptance of all of Lynn's views, we must gratefully recognize her deep influence on biology and her exemplary role in the establishment of earth system science.

This leaves us with the question of what the use can be of postulating the new symbiotic worldview. A chorus of philosophers and opinion makers is claiming that since the ideologies of the twentieth century melted away, we're in need of a new "big story." This long view is exactly what earth system science provides. This is borne out by the mounting success of "Big History," a movement inspired by David Christian,[14] Fred Spier,[15] and many others. In several universities, Big History courses "from the big bang to globalization" are being implemented as general background information for all of their students (see http://www.ibhanet.org). We may regard earth system science as a critical component of Big History, of special importance because it immediately relates to our own position in nature. The vision inspired by earth system science may seem scary at first, but it turns out to be a liberating experience to leave our cocoons, become aware of the immeasurable depth of our roots, and reorient our lives in agreement with the real world.

Maybe the greatest merit of the symbiotic worldview is its potential to clarify our mind and to give us a new orientation in life. It draws the line between practices and mental representations that frustrate a symbiotic relationship between civilization and the system earth and those promoting such an arrangement. For example, politicians who exclusively see the betterment of humanity, and not the differentiation of the earth, as the ultimate goal of their efforts, are one-sided and will do more harm than good, however noble their intentions. They remain locked up in the human cocoon, unable to place their policies in the wider context of global dynamics. Similarly, the science of economy will never reach maturation without the top-down systems approach that regards their subject as a global geological process or force. Good surfing takes practice, and to suppress the idea that we are the masters of the earth requires a great deal of discipline. Let us remember that practice and discipline are fundamental to the entire civilization process.

On the other hand, we should not fatuously engage in false modesty as to the role of humanity within this ratcheting planet. The symbiotic worldview has more to offer than anxiety and humiliation, as it also endows us with a new and unexpected dignity. After all, when the *Apollo 8* astronauts observed Earth from deep space, it was Earth seeing itself for the first time in 4.6 billion years—through our eyes. When science seduces the earth to reveal its intimate workings, it is the earth that discovers itself—through our effort. And we humans are the principal players in the civilizing process, the latest stage of global differentiation. When the worst comes to the worst, we shall need this minimal self-esteem to follow the example of the hero fisherman in the maelstrom who found an attitude of detachment and fascination and abandoned the ship that had been his only protection so far. Like him, we should not hesitate to throw ourselves right into the boisterous waters.

Peter Westbroek, a Dutch geologist and professor emeritus of geophysiology at Leiden University in the Netherlands, is author of Life as a Geological Force. *He may be found talking about Wagner and the earth on a TEDx talk here: http://www.youtube.com/watch?v=s64x5PuslBA.*

Rebel, Teacher, Neighbor, Friend

Gaiadelic:
Lynn Sagan and LSD

RICH DOYLE

Lynn Margulis had a profound impact on any answer to the question that makes up the title of one of the finest books of late twentieth-century popular science, coauthored with her son Dorion Sagan: What Is Life? While the neo-Darwinian synthesis answered Schrödinger's famous and eponymous query with the cartoonish absurdity of self-propagating lumbering robots enslaved to the proliferation of their own genetic information, Margulis and Sagan respond with a more subtle and, yes, mind-expanding account of living systems, one that recognizes a sublime continuity among all forms of life, as each living system itself becomes an aspect of an interdependent network much larger in scale and frame of reference than any of the participants. While molecular reductionism offered the fruitful and, within its frame of reference, accurate portrayal of living systems as mechanical systems that can be subdivided and understood by consciousness ad infinitum, Margulis and Sagan reintegrate this intensely localized view of life—DNA as the microcosm—within the resonant fabric of the whole, one that includes the consciousness of the observer.

This inclination toward holarchy—she also liked the term "holonomy," a derivation of Koestler's term meaning "nested autonomy"—is a fractal characteristic of Margulis's immense contributions to biology: each part of her itinerary replicates the whole, a propensity toward the whole. From her early and persistent efforts to incorporate symbiosis into our models of evolution to her collaborative and courageous efforts to model the homeostasis of earth itself as Gaia, Margulis explored the tensions and relations between part and whole in living systems. It is perhaps less evident that this holarchic view of living systems—in which the whole as well as the parts plays a role in all living systems of interdependent beings—also suggests a theory of mind that is at least as audacious as endosymbiogenesis or Gaia theory, and like these models it has the potential to radically alter our models of what it means to be conscious, as well as alive.

Margulis's published theory of mind predates her more famous 1967 contribution to a symbiotic theory of evolution, "On the Origin of Mitosing Cells," and so, too, is this theory of mind bundled necessarily with a model of truth. It is this model with which we will begin. Responding to an article by psychologist Joe Adams, "Psychosis: Experimental and Real," Lynn Sagan, then a twenty-five-year-old freshly minted PhD from University of California, Berkeley, and young faculty member at Boston University, published "An Open Letter to Mr. Joe K. Adams" in the fledgling journal of record for the newly intensified research into the alteration of consciousness, *Psychedelic Review.*

> As a "socially unacceptable truth-teller" in all subculture normalizations except (as I recently discovered to my great joy) the eclectic and tolerant one in and around Harvard Square, I responded to your article in the *Psychedelic Review* with the kind of zeal Herman Cortes must have felt after discovering that the Mayan girl Malinche could talk to the Tarascan natives for him.[1]

A scholarly and scientific forum for the inquiry into the "exploration of consciousness," the *Psychedelic Review* published eleven volumes between 1963 and 1971 with contributions by biologist Julian Huxley, psychologists Timothy Leary and Ralph Metzner, theologians Walter Pahnke and Alan Watts, and BBC science writer and mystic Gerald Heard—a veritable who's who of early 1960s psychedelic science. In her two-and-a-half-page response to "Psychosis: Experimental and Real," Lynn Sagan articulates a model of truth and consciousness as vectors for a decompartmentalization of knowledge and experience:

> Our university classrooms are filled with your "compartmentalists," camouflaged with pedantic verbiage and fancy formality. I have long been their antagonist, and have learned as well as suffered. As reference, I cite my own experience with the compartmentalist designated by the venerable titles "biochemist" and "biophysicist," as well as with LSD-25. Are those credentials enough to comment on your hypothesis?[2]

Adams's hypothesis suggested that psychedelics can trigger psychotic episodes occasioned by "a sudden collapse of boundaries between two or more cognitive structures previously kept separated from each other within that particular individual's total set of cognitive structures."[3] Adams also suggested that while the result can sometimes be labeled "psychotic," these episodes also enable a more rather than less accurate perception of reality through the dissolution of categories that model rather than reveal reality: "All the psychedelic or 'mind-manifesting' drugs attack the defense of compartmentalization and thus make it possible for an individual to see through some of the absurdities, including status systems, of his own behavior, and of his own culture and groups-of-reference."[4]

A young Lynn Sagan agrees and wonders, perhaps as a scientist, at the social rejection of increased reality. She writes, "I really do agree with your contention that the drug attacks defense mechanisms built up carefully to conceal the truth of our direct sensory perceptions. One would a priori imagine, however, that a drug which forced us to see the world as it is would be welcomed. Why, then, is the entire 'consciousness-expanding' drug movement confronted with enormous hostility?"[5]

But before answering her own question, Sagan takes note of the scientific stakes: "The excitement here arises from our present position: we are probably on the threshold of a physical basis of consciousness. Perhaps our times are analogous to those at the beginning of the century, which culminated in today's clear concept of the physical basis of heredity."[6]

For Sagan, here, science has become recursive: with the material tools of psychedelic compounds and plants, it becomes possible to scientifically study the very basis of science itself—consciousness. While subjectivity had been separated from the domain of scientific investigation at least since the Royal Society's insistence on a "modest witness" who observed but did not affect experimental inquiry, psychedelics promised to rigorously reintegrate consciousness into scientific inquiry by including an observer who observed herself under the conditions of "psychonautic" exploration. Sagan's analogy to the discovery of DNA—"the physical basis of heredity"—suggests that for her as well as other researchers of the period, psychedelic science augured new paradigms of mind. But as with past emerging paradigms—the Copernican and Darwinian worldviews, for example—new maps of reality also threaten existing worldviews that perhaps are not true so much as they are claimed to be true. In this context "truth" is less the characteristic of a proposition than an activity—an ongoing "antagonism" to the categorization and compartmentalization of the world, the practice of insisting on the "directly sensed." Sagan notes that this tension between what is true and what is claimed to be true enables the evolution of human beings as social animals:

> In addition, then, to accurate individual perceptions of the external world, man must contain within his nervous system the profound tolerance for ambiguity between what is directly sensed and what is claimed to be "true" by others. If one could not perpetuate mystiques such as "Communists are evil and out to destroy our American society," or, "Our actions are accountable to an invisible God," for example, thousands of people would not be mobilizable into actions which *per se* are atrocious and unthinkable.[7]

If, for Sagan, "true" is then grammatically to be understood as a verb, then "truth" is also first person: it emerges from the "directly sensed" experience of that disjunction between the claim of "true" and the psychedelic or "mind manifesting" experience of watching the categories of official reality become . . . categories. Again, this "psychonautic" investigation is by necessity a first-person experience: "I cite my own experience with the compartmentalist designated by the venerable titles 'biochemist' and 'biophysicist,' as well as with LSD-25."

At first glance, one might expect the emerging psychedelic theory of mind from 1963 to be a reductionist one, an inquiry that identifies the biochemical nature of mind with chemical activities—such as the serotonin agonism of LSD-25—on the molecular scale. Yet the chemistry, for Sagan, reveals not only the obvious role of chemical interactions in the alteration of consciousness but, more crucially, the capacity of consciousness to study itself: "The chemistry eloquently testifies to the amenability of man's soul to his own researches."[8] This suggests not only that consciousness can be altered by compounds and plants but, perplexingly, that it can transcend such altered consciousness and act as an observer on its own systemic alteration. Hence, in the thick of a radically materialist investigation into the nature of mind, we find hints of a theory of mind in which an observer's consciousness somehow transcends its own material alteration. This is an experience had by many psychonauts, who learned that they could both alter their consciousness and somehow recursively witness this altered consciousness. Swiss chemist Albert Hofmann, who synthesized the drug LSD-25 in 1938, discovering its effects in 1943, described his own first deliberate ingestion of the drug in similar terms in his autobiography, *LSD: My Problem Child*:

> This self-experiment showed that LSD-25 behaved as a psychoactive substance with extraordinary properties and potency. There was to my knowledge no other known substance that evoked such profound psychic effects in such extremely low doses, that caused such dramatic changes in human consciousness and our experience of the inner and outer world.
>
> What seemed even more significant was that I could remember the experience of LSD inebriation in every detail. This could only mean that the conscious recording function was not interrupted, even in the climax of the LSD experience, despite the profound breakdown of the normal world view. For the entire duration of the experiment, I had even been aware of participating in an experiment, but despite this recognition of my condition, I could not, with every exertion of my will, shake off the LSD world. Everything was experienced as completely real, as alarming reality; alarming, because the picture of the other, familiar everyday reality was still fully preserved in the memory for comparison.[9]

Hence the promise of psychedelics was not only that they represented precise probes for systematically altering human consciousness, but that in the very alteration of consciousness they revealed a layer of human consciousness capable of integrating both the experience of altered consciousness and "familiar everyday reality." The experience of this "conscious recording function" pointed to a larger-scale experience in which the localized ego was a manifestion of the whole. A participant in a 2006 study at Johns Hopkins University on the effects of psilocybin continued this refrain: "To cease to 'BE,' as I understand it, was not frightening. It was safe and much greater than I have words for or understanding of. Whatever is larger than the state of being is what was holding me."[10]

This pattern of looking past the compartmentalization of reality to the larger-scale integration of the part and the whole—such as a scale "larger than the state of being" capable of integrating both being and nonbeing—repeats fractally through Margulis's epic scientific journey: the "greatest single evolutionary discontinuity to be found in the present-day living world,"[11] the prokaryotes and eukaryotes, become integrated by the oxygenated atmosphere via a symbiotic vector leveraging not the natural selective separation of organisms but their evolutionary and thermodynamic integration.

> It is suggested that the first step in the origin of eukaryotes from prokaryotes was related to survival in the new oxygen-containing atmosphere: an aerobic prokaryotic microbe (i.e., the protomitochondrion) was ingested into the cytoplasm of a heterotrophic anaerobe. This endosymbiosis became obligate and resulted in the evolution of the first aerobic amitotic amoeboid organisms.[12]

Lynn Margulis and Dorion Sagan will later define the continuum of consciousness to include this larger scale, as in their treatment of "gradient perception" in *Acquiring Genomes*, where "mind" emerges from the tug of that scale integrating being and nonbeing, the cosmic imperatives of thermodynamics: "This new thermodynamics (sometimes called homeodynamics) lets us begin to glimpse the path from matter (gradient breakdown) to mind (gradient perception)—from energetic to informational 'self' organization."[13]

So, too, does the historically audacious claim for Gaia emerge from this holarchic integration whereby the whole—in this case, the biosphere—acts as part of the system:

> The total ensemble of living organisms which constitute the biosphere can act as a single entity to regulate chemical composition, surface pH and possibly also climate. The notion of the biosphere as an active adaptive control system able to maintain the Earth in homeostasis we are calling the "Gaia" hypothesis, Lovelock (1972). Hence forward the word Gaia will be used to describe the biosphere and all of those parts of the Earth with which it actively interacts to form the hypothetical new entity with properties that could not be predicted from the sum of its parts.[14]

Beginning from Lynn Sagan's testimony concerning her antagonism toward compartmentalization, then, we can perceive the activity of her own consciousness as a vector of continual "decompartmentalization," a search engine for holarchy. Following her own logic and testimony, we might note that this vector emerges not only from the local instance of mind whose borders were found at the boundaries of Margulis's skin, now itself subject to the unmistakable evidence of the aforementioned cosmic imperatives of thermodynamics. Instead, following the path of her thinking, we perceive something larger scale, perhaps not only "more" accurate but different in kind: we—and by "we" I mean not just the "star stuff" Carl Sagan spoke about but also the living microbes beneath our feet and the planetary being of which we're a part—are a way for the (micro)cosmos to know itself.

Richard Doyle is liberal arts research professor of English and information sciences and technology at Penn State University and is the author of a trilogy of books on information and the life sciences. The latest, Darwin's Pharmacy: Sex, Plants, and the Evolution of the Noösphere, *was published by the University of Washington Press in 2011. Professor Doyle is a fellow at the Hybrid Reality Institute and a Distinguished International Fellow of the London Graduate School.*

Two Hit, Three Down—

The Biggest Lie: David Ray Griffin's Work Exposing 9/11

LYNN MARGULIS

I comment here on the nanotechnology aspect of Jerry Mazza's masterful review (*Rock Creek Free Press*, January 2010, p. 6) of David Ray Griffin's extraordinary 2009 book, *The Mysterious Collapse of World Trade Center 7: Why the Final Official Report about 9/11 Is Unscientific and False*.

By the time we (Dorion Sagan—my eldest son and Sciencewriters partner—and I) met David Griffin in 2003 in his native habitat at the Center for Process Studies (which is on the campus of the Claremont School of Theology in Southern California), he had written over two dozen books, none of which I had ever read or even heard of. We immensely enjoyed a three-day scientific-philosophical meeting on the Darwinian-evolutionary view of life that had been organized by Griffin's sage mentor, the sweet-tempered but razor-sharp octogenarian professor emeritus and director, John B. Cobb Jr. The results of this fascinating meeting have since been published.[1]

At that meeting, Griffin's talk was sober, academic, competent, scholarly—and entirely new to me: Christian theology in a much broader philosophical context than any to which I had ever been exposed. The science-friendly philosophical outlook Griffin espoused apparently was developed by Alfred North Whitehead (1861–1947), the English mathematician-philosopher who became a Harvard professor, or by Cobb. Why would I have known anything about this theological-philosophical work? My own expertise after all is in protoctist and organellar genetics. With close colleagues I reconstruct the origin and evolution of nucleated cells in the Proterozoic eon. Where I understood DRG's talk at all, he made clear to me his honesty. Truth, especially scientifically/empirically established truth, seemed intrinsic to his Christianity. As a typical agnostic scientist overtly critical of organized, and even disorganized, religion, I was surprised by the scientific commitment to approaching empirical truth in a religious context rather than the usual authority-pleasing consensus.

Griffin went on to elaborate that although as a Whiteheadian he embraces the methods and results of science, he is critical of the entire international scientific enterprise as generally practiced today. Not only do scientists extrapolate their intrinsically specialized empirical knowledge into intellectual territory where it does not belong, but they don't heed Whitehead's recognition that, after all, scientists—like all people—have an emotional life and an inner spirit. The acclaimed "objectivity of science," on close scholarly inspection (by, say, Cobb, Griffin, and other Whiteheadians), often translates into tribalism, jingoism, naïveté, denial of obvious truths, uncritical service to the state, and other forms of profoundly dangerous ignorance. Scientists, as do all groups of people, share unstated philosophical assumptions. For example, they not only have faith in the consistency and "knowability" of the real world, but they often assume that the concrete particulars of the world are adequately described by the abstractions that have proved useful for limited purposes in their own disciplines—an assumption that Whitehead called the "fallacy of misplaced concreteness."

On the return trip east from this meeting, I read, nonstop, DRG's seminal book on 9/11, *The New Pearl Harbor*. Since then I have watched the "aging theologian"—his self-description, as quipped in the excellent videographed lecture "9/11 and Nationalist Faith" (or was it "9/11: Let's Get Empirical"?—both are highly recommended)—metamorphose from a compelling, careful scholar to a brave and extraordinary orator (but still careful scholar). Griffin has become a superb politician with a single agenda item: we must reopen public inquiry into the events of September 11, 2001, especially the collapse of the World Trade Center.

I had personal reasons to be interested in this issue: I had watched a member of our Geosciences Department (University of Massachusetts-Amherst) realize that morning that his beloved brother was on the doomed Boston-to-LA "hijacked" plane. Two of my sons and both of their mates were on Manhattan Island on 9/11. My grandson Tonio Sagan, the Problemaddicts hiphop lyricist and leader, was released from his Springfield probationary status to me that evening. Like everyone else close to the action, my global consciousness was instantly and permanently altered on that date. But I happily remained an academic evolutionist. So, I ask here, what happened to the "aging theologian" to cause such a radical shift in his personal and professional life: from ivory tower scholar expert in theology and philosophy to detective, orator, and political activist?

I will not try to answer this question but will simply state: Griffin has become a scientist, in my view, and even more a science educator. He has undertaken the search for the solution to a relatively trivial scientific problem and has found it in the literature and through discussions with experts. Along with solving the scientific problem, however, he has burdened his life with a colossal science education problem.

In spite of his authorship of eight excellent books on the subject, he is not winning the education skirmish. This is not surprising, because science education battles—I can tell you this from chronic painful experience—are far more intrinsically difficult to win than those of mere science. I illustrate this point with regard to the destruction of the World Trade Center.

The scientific problem:

Why did three World Trade Center buildings (1, 2, and 7) collapse on 9/11, after two (and only two) of them were hit by "hijacked" airplanes? The scientific answer:

Because all three buildings were destroyed by carefully planned, orchestrated, and executed controlled demolition. Ignited by incendiaries (such as thermate) and high explosives (including nanothermite), the steel columns were selectively melted in a brilliantly timed controlled demolition. Two 110-story buildings (towers 1 and 2), plus one 47-floor building (WTC 7), were induced to collapse at gravitationally accelerated rates in an operation planned and carried out by insiders. The apparent hijacking of airliners and the crashing of them into the Twin Towers were intrinsic parts of the operation, which together provided a basis for claiming that the buildings were brought down by Muslim terrorists. The buildings' steel columns, which would have provided irrefutable physical evidence of the use of explosives, were quickly removed from the scene of the crime.

Impeccable logic and addiction to reading inspire this truth-seeker. With his practiced scientific mind (he has organized professional philosophy of science conferences and published several books on this theme), coupled with investigation of evidence from not only us scientists but from witnesses, documentary filmmakers, scholars, architects, engineers, and others, he concludes that the virtual free-fall collapse of the three (not two) World Trade Center buildings in 2001 was a premeditated, exquisitely executed operation. He recognized this "tragic publicity stunt" (my claim) was likely intended to provide the "new Pearl Harbor"[2] desired by radical neoconservatives, some of whom had become members of the Bush administration (see chapter 6 of Griffin's 2006 book, *Christian Faith and the Truth behind 9/11)*.[3]

The far more difficult science education problem:

The persistent problem is how to wake up public awareness, especially in the global scientifically literate public, of the overwhelming evidence that the three buildings collapsed by controlled demolition. (Much has been published in peer-reviewed scientific journals; see chapter 4 of *The Mysterious Collapse*).[4] We, on the basis of hard evidence, must conclude that the petroleum fires related to the aircraft crashes were irrelevant (except perhaps as a cover story). We citizens of earth within and beyond the boundaries of the United States who demand detailed evidence for extraordinary claims agree with Griffin: the rapid destruction of New York skyscrapers on September 11, 2001, was planned and executed by people inside the US government.

Griffin's eight books about 9/11 are his call to his kind of truly patriotic action. They show in appropriate detail, accurately documented, that the official government conspiracy theory cannot be correct journalistically, scientifically, and morally. Muslim airline hijackers, in short, never triggered the collapse of high-rise steel-framed buildings at gravitational acceleration into neat removable pieces. They didn't remove the remaining steel girders before they could be studied as evidence for a huge crime. They were not described in telephone calls by passengers and crew members from the four airliners—all the evidence for al-Qaeda hijackers on the planes dissipates under close inspection (for example, even the FBI now admits that the reported cell phone calls from 25,000 to 40,000 feet never happened).

And to me the most compelling and obviously incorrect accusation is that Muslim hijackers caused the pulverization of cement high-rise office buildings into tons of dust that contain crystalline thermate and other minute metallic particles not found in the usual charred remains of fire rubble. Minute iron-aluminum-molybdenum-rich spheres, steel perforated with swiss-cheese–type holes, and large quantities of unreacted nanothermite are not components of petroleum office fires. Besides the fact that building fire temperatures, even if fed by jet fuel, could not have risen beyond 1,800°F, and hence they would be nowhere close to the 2,800°F needed to melt iron (or for molybdenum, which melts at 4,753°F). The facile appeal to the presence of gypsum (calcium sulfate) in the office wallboard fails to explain why sulfur was found in the intergranular structure of pieces of steel. (There was no detection of calcium!) *The New York Times*, in a rare example of honest reporting about the WTC collapses, called this "the deepest mystery uncovered in the investigation".

Nor could "Muslim terrorists" have accessed and then planted in these buildings huge quantities of nanothermite. This recently developed high explosive was developed mainly in secrecy by professional scientists and engineers who enjoy government grant support for "nanotechnology" by the military. Significant quantities of red-gray crystals of nanothermite have been found in several independently collected samples of WTC dust studied by a team headed by physicist Steven Jones, formerly of Brigham Young University. Niels Harrit, a University of Copenhagen professor of chemistry who specializes in nanochemistry, is the first author of a peer-reviewed paper reporting this team's results.

The two mutually exclusive "9/11 conspiracy theories," the patently and nefariously absurd tale our government imposes on us and the true, criminal story yet to be entirely brought to light, deserve the attention of all literate people. Remember: only two airplanes struck, but three buildings collapsed at free-fall velocities on that same day. Begin with Mazza's review and Griffin's book that detail the nanothermite scientific studies. Examine the government's reluctant late admission that WTC 7 came down in absolute free fall for over two seconds to realize this is a scientific impossibility unless all the steel columns were partitioned into neat, removable pieces by explosives. Find out what happened to two men, Hess and Jennings, trapped inside WTC 7 in its abortive explosion before noon. Truth here, as David Ray Griffin tries to tell it, is (at least to me) stranger and far more dramatic than even the best fiction.

No Subject Too Sacred

JOANNA BYBEE

Some people say that the greatest benefit of a liberal arts education is not learning a specific subject matter, but rather learning to learn. This may sound pat and obvious, except that learning, and learning to think for oneself, is actually a radical act. I was privileged to work in the Margulis lab during my junior and senior year as an undergraduate student at UMass-Amherst in 2006 and 2007. In addition to sharing her encyclopedic knowledge of evolutionary biology, genetics, microbial ecology, and geochemistry, Lynn taught me science: a way of knowing by observing the evidence, regardless of how it threatens the entrenched interests of those in power.

Under the supervision of Dr. Michael Dolan, I studied amitochondriate protists inside the wood-feeding termites *Cryptotermes cavifrons* and *Neotermes mona*. These single-celled protists living in the hindgut of termites are important because, without any mitochondria, they testify to the chemical environment of the early Earth, before there was free oxygen in the atmosphere. I studied such hindgut protists as models for early eukaryotic life and wrote my thesis on the controversies surrounding the taxonomy of the wood-feeding cockroach, *Cryptocercus*.

The phylogenetic placement of the genus *Cryptocercus* of the modern-day cockroaches in the order Blattaria spurred debate as to whether it should be considered a sister group to the termites (order Isoptera) or if rather it belonged as a sister group to the mantids (order Mantodea).

Looking at the extensive genetic variation between the known *Cryptocercus* species that inhabit the temperate forests of the Palearctic and Neararctic in the United States, Russia, and China, we can see that the species diverged from each other tens of millions of years ago. More fascinating to me was how entomologic authorities couldn't agree about where to place the wood-feeding cockroach on the tree of life even with molecular and morphological data. At my thesis defense, I outlined the different viewpoints held by the experts in the field and when asked what I thought, I didn't feel comfortable taking a side.

To Lynn, the cell was the basic unit of life and communities are the unit of selection: "You cannot understand anything in biology without recognizing the community nature of all life whether they're loose bacterial communities or they are very tight ones that we recognize as cells or whether they're even tighter ones that we recognize as animals or plants."[1] She admired Charles Darwin for his insight that life began with a common ancestor, but she argued the topology of evolution from there was more of a web than a tree. Her arguments for the importance of symbiosis, based on decades of immersion in the literature, laboratory, and field investigations, once dismissed, have been confirmed by genetic evidence and are now taught in textbooks.

For Lynn, books had their place but were also worthy of scrutiny. I remember her quoting naturalist Louis Agassiz, "Study nature, not books," while she showed me a distorted diagram in a textbook meant to depict the cellular structure of *Euglena*. The undulipodia was misnamed a flagellum and drawn so that it bisected the cell, when in nature it can only be seen on the anterior of the organism. She told me the mistake had been copied throughout the scientific literature. Over the years, she showed me microscopic images of mitochondria as well as other endosymbionts that couldn't be found in the cytological diagrams of biology texts.

Every aspect of Lynn's life echoed an empirical scientific stance and a search for the truth. Despite criticism from her peers, she held fast to her observations and understanding of symbiogenesis. "On the Origin of Mitosing Cells" was rejected fifteen times and many of her grant applications turned down, with one response being, as reported in the UK newspaper *The Telegraph*, "Your research is crap. Don't ever bother to apply again."[2]

In a BBC interview, Dorion Sagan spoke of Lynn's courage:

> I think she had a sort of a congenital lack of fear. Her symbiosis theory was made as a young woman and so it emboldened her and it bolstered her in continuing to press the envelope and what she thought was right and true based on the evidence. She looked at things herself and she was often right where other people were wrong. Galileo said something very similar about a single scientist in his basement working diligently [and his results] being worth more than the received consensus of all the others.[3]

Asked once why there was a scientific backlash against her symbiogenesis theory, she replied:

> I'll tell you why, there are a lot of reasons. Some of them are just our language and some of them took me years to realize because I was taught the same language as everybody else. It's a representation of the paradigm just like reciprocal altruism and fitness and the parental investment of male children are indicative of this neo-Darwinist paradigm. Neo-Darwinists are people who emphasize individual differences, selection of individuals, which make us people nuts because an individual can't even reproduce, right? It's a system of looking at the world and it's reflecting Victorianism, nineteenth-century capitalism versus communism.

The interview continued: "What is it about symbiosis that people don't like, [do] they think it's this leftist conspiracy?" Lynn replied, "It's female. [It] is cooperative and noncompetitive. Doug Caldwell did an analysis of the *Origin of Species* and the numbers of times that competition, struggle, fight to death, these words, it's something like 10,000 to 1." She continued, "I think that monarchy and monotheism that is deep within our culture, your culture more than mine, I mean the monarchy part, that when they talk about monophyly in biology which means one single ancestor, it's just an extrapolation from monarchy and monotheism."[4] Not only did she disagree with the conventional model of evolution—diversification of species resulting from the accumulation of beneficial genetic mutations—that neo-Darwinists such as Richard Dawkins champion, she abhorred the idea that an individual's genome, let alone individual genes, were considered the unit of evolution. For her it was cells, all the way down. Cells were the minimal unit of evolution. They speciated and diverged, but they also merged, as did organisms made of them, and populations and communities.

The first field experience listed on her CV occurs in 1956, when she was just sixteen and studied in Tepoztlan, Morelos, Mexico, with anthropologist Dr. Oscar Lewis. I learned of her interest in Richard Evans Schultes, of his discovery of psychedelic plants in Mexico, when I heard her interview him on one of her digital interactive learning videos she taught in her environmental evolution class.[5] The ayahuasca- and peyote-inspired artwork in her house perhaps reflects the lasting influence of the formative fieldwork she conducted in that Mexican village where, registered as a student at the University of Illinois, she studied with "the modern doctor and the curandero," that is, shamanic healers. Intriguingly, the swirling multicolored forms in the artworks reminded me of the symbiotic bacteria that were her lifelong passion.

In my experience with Lynn, I learned that no subject was too sacred to go untouched and unquestioned. She thought the acquired immune deficiency syndrome (AIDS) was not caused by the human immunodeficiency virus (HIV). She was critical of the Cuban embargo; while at Boston University, Lynn conducted fieldwork in Cuba and helped to create an organization called the North American/Cuban Scientific Exchange in an effort to form collaborations with Cuban scientists. She thought the eukaryotic undulipodium, commonly referred to as flagellum, evolved from an ancient spirochete symbiosis: "Do you want to believe that your sperm tails come from some spirochetes? Most men, most evolutionary biologists, don't. When they understand what I'm saying, they don't like it."[6]

Rather than being fazed by opposition when she believed she was right, Lynn redoubled her efforts. One journalist said she "defined herself by oppositional science,"[7] but that's an oversimplification. Lynn was persuaded by physical evidence and was dissuaded by strident voices, attempts at intellectual intimidation, and appeals to authority.

One day, late in the spring semester of 2006, during her tenure as president of the prestigious research society Sigma Xi and shortly after my induction, Lynn handed me a DVD. The documentary called into question the commissioned report of 9/11 based on contradicting news reports and other physical evidence.

My first thought was, "Why would she believe this stuff?" What Lynn didn't know about me at the time was that my twin brother was serving his first tour in Fallujah, Iraq, as a private in the marines. I spent hours on the phone with him over the course of his tour, as he told me stories of IEDs that shook buildings, snipers that almost took out his friend on a rooftop, and something he did not tell me but I later found out: a suicide bomber made his way past him as he stood guard at one of the many checkpoints, an event that eventually led to the death of a fellow soldier in his unit.

Perhaps seeing my skepticism, Lynn looked at me and smiled knowingly in the doorway of her office in Morrill Science Center. She told me a story about the USS *Maine*, suggesting that the explosion that killed most of the crew on February 18, 1898, while the ship was docked in Havana harbor may have been a false-flag operation, creating the pretext for the Spanish-American War.

Lynn urged me to question: if a false-flag operation could have conceivably happened in the past, what makes it an impossibility today? She always encouraged me to look directly to the evidence, whatever the official story might be.

In her essay "Two Hit, Three Down–The Biggest Lie," Lynn alleged that 9/11 was a false-flag operation, probably participated in by radical neoconservatives in the US government who later joined the Bush administration. Despite my discomfort after reading her essay, I began to wonder, How could building 7 be completely demolished in seconds by the debris from one of the Twin Towers while buildings 4, 5, and 6, which sustained more damage, still maintained their structure? How can a single-point failure of one column be to blame if kerosene type A jet fuel cannot melt steel?[8] What of the crystalline nanothermite, a metastable intermolecular composite known to be used in military explosives that were claimed to have been found in the debris by Niels Harrit, Steven E. Jones, and colleagues that was published in *Open Chemical Physics Journal*?[9] When I brought this up with my family, I was shut down. I quickly realized that the official story of September 11 was not a topic open for discussion. But I wondered: why is it not okay to make rational inquiries concerning matters of great political importance?

In an interview that appears on 911blogger.com, she stated, "Evidence was either destroyed or removed systematically, extremely close to the time that the crime had occurred. This is not only unscientific, it's illegal."[10] She lauded the work of David Ray Griffin, whom she met at a conference and who inspired her to read his first book about 9/11, *The New Pearl Harbor*. In the essay, she continues, "From there I went on to read his even more disturbing account of the bogus 9/11 Commission Report, *The 9/11 Commission Report: Omissions and Distortions*, which provides overwhelming evidence that the official story is contradictory, incomplete, and unbelievable."

In the beginning of her essay Margulis recounts her admiration for Griffin's Whiteheadian criticism of the entire international scientific community's "fallacy of misplaced concreteness." She felt an unnecessary disconnect within the community due to a language barrier caused by jargon and sometimes the reification of these terms. She would attest to the claims of "objectivity of science" in some of her peers as nothing more than, "tribalism, jingoism, naïveté, denial of obvious truths, uncritical service to the state, and other forms of profoundly dangerous ignorance."

In an interview, Lynn once questioned: "Why are people refractory? Why are people resistant? Why don't people see the evidence when it's out there? And the answer is that for people it is much more important to be approved by their group and belong to their social group than it is some arbitrary interest in the truth or scientific truth."

As an undergraduate studying the controversies concerning *Cryptocercus,* I learned that there could be many interpretations of the same data set, but more important, we need to look at what's included and what's missing. There is such a thing as superfluous data, variables that have nothing to do with the hypothesis in question, and at the same time you must always keep in mind that you don't know the whole picture. I began to understand Occam's razor, the law of parsimony that calls for selecting the hypothesis that contains the fewest assumptions, and a dictum sometimes referred to as Einstein's razor: "Make things as simple as possible but no simpler."

My journey with Lynn deep into the microcosm and out over the Gaian macrocosm challenged many of my own beliefs, including scientific and political paradigms that, until I met Lynn, I would never have dreamed of questioning. She taught me that if errors concerning something as seemingly objective as natural history can be perpetuated in textbooks, why should we believe, without question, what we are told of politics and war, where people are that much more motivated to trick, lie, and deceive? Science is about the empirical search for truth in an objective manner and being receptive to the answers, no matter how uncomfortable they may make us.

Joanna Bybee is a biologist working in the biotechnology industry in the greater Boston area.

Next to Emily Dickinson

TERRY Y. ALLEN

Speaking in a live broadcast at the Cambridge Forum less than a year before her death, Lynn Margulis said, "Emily Dickinson talks to me all the time. She is my neighbor. She is ironic. She exposes pretensions. She is a botanist. She is my favorite poet. And she is arguably the greatest English-language poet."

Margulis and the deceased poet (1830–1886) were indeed neighbors. In 1988, when she accepted an appointment at the University of Massachusetts, Margulis moved from West Newton outside Boston to a large white Victorian house at 20 Triangle Street in Amherst. She soon discovered that her property adjoined the venerable three-acre Dickinson site, now the Emily Dickinson Museum. The shared boundary is at the northeastern edge of the Dickinson property. In that quiet section of the family compound, the green-thumbed poet—always dressed in white and usually kneeling on a red blanket, according to her contemporaries—cultivated her flowers. Today the Emily Dickinson Museum maintains the charming small garden, with stone steps and a bench, on a portion of lawn that slopes down to Triangle Street. The garden can be seen from the Margulis house.

The nearly 1,800 lyric poems that were discovered and published after the poet's death at age fifty-five reveal a powerfully original, probing mind. Dickinson received a first-rate science education in the 1840s at Amherst Academy and at Mount Holyoke Female Seminary (which became Mount Holyoke College in 1893). Always an excellent student, Dickinson formally studied chemistry, geology, and botany, among many other subjects. In a letter home to her brother, Austin, she mentioned that in the Mount Holyoke laboratory she had been engaged in the study of "Sulfuric Acid !!!!!" Yes, the poet used five exclamation points.

In 1855, years after her return from Mount Holyoke, Dickinson's father, Edward, had a small greenhouse built off the dining room of the Homestead, as the mansion was known. This space became Emily's domain, a place where she grew native and exotic plants that flourished under her tender care even in the depths of the New England winter. With a straight pin, Dickinson sometimes affixed a flower or blossom to a poem before sending it off to one of her favored correspondents.

Not long after her arrival at UMass, Margulis's graduate students gave her a book of Dickinson's poems. For years, as she studied, memorized, and recited impressive chunks of Dickinson's work, Margulis developed a kinship with the poet who had died more than a century before they became "neighbors." She linked the poet's working genius "to my increased awareness of the budding scientific field of biosemiotics."[1] Margulis wrote that Dickinson's poetry is alive to the cycles and mysteries of the natural world, bodily sensation, and the significance of symbols, signifiers, and phrases.

Nearly two decades after Margulis arrived at UMass, Professor James W. Walker, the UMass botanist and evolutionist who had recruited her, asked a favor. Would she be willing to approach the well-known neurologist Oliver Sacks about delivering the annual William E. Mahoney Lecture? This lectureship was funded by two chemistry graduates who were also planning to endow a new integrated science program at their alma mater. In attempting to scale the impenetrable wall that surrounded the reclusive neurologist, Margulis appealed to her friend Roald Hoffmann for help. Hoffmann, a professor at Cornell University and a Nobel laureate in chemistry, is also a poet. In the end, it was he who gave the 2007 Mahoney Lecture. In accepting the invitation, Hoffmann made one condition: he must see Emily Dickinson's house.

The Amherst-based scholar Ruth Owen Jones guided Margulis and Hoffmann on a tour of the museum. At the time, Jones was doing research for her biography of William Smith Clark (1826–1886)—eminent man of science, Civil War hero, founder of universities, and

Amherst resident—following her publication of a controversial paper titled "Neighbor and Friend and Bridegroom: William Smith Clark as Emily Dickinson's Master Figure." In three extant "Master letters," discovered in the poet's bedroom after her death, Dickinson artfully and abjectly addresses an unknown person whom she appears to love deeply. While Dickinson never married, she did leave behind a trove of erotic poetry, some of it composed during the period in her late twenties and early thirties when she wrote the extraordinary Master letters. Many scholars, like Jones, have attempted to pin down the identity of a lover.

Margulis endorsed Jones's Clark-as-Master thesis and urged her to complete the biography. In addition, Margulis tried to enlist the filmmaker Terrence Malick, the director of *Days of Heaven* and *Tree of Life*, for which Margulis served as a consultant, to make a film based on the Clark thesis.

Not long after the Mahoney Lecture, Margulis received a letter from Roald Hoffmann that introduced her to the scholarship of the late Swiss-born Hans Werner Lüscher (1901–1991). An emigré, Lüscher earned his living in Los Angeles by day as a carpenter; nights were for his philosophical writings and essays. After he had translated some of Dickinson's poems into German, Lüscher believed he had made a sensational discovery: the poet, he believed, had developed a secret, double language in her poetry for recording her hidden sexual life. Dickinson had elected, in her own famous formulation, to "tell all the truth but tell it slant."

In all, Lüscher analyzed 1,154 Dickinson poems and created a glossary of 257 major symbols. Lüscher wrote that he "could not get rid of an impression that a nimbus of symbolic double meanings hovered around a great number of nouns and about some verbs, adverbs, and adjectives likewise." Having, as he thought, broken the poet's private code, he reasoned that no virgin could have written as Dickinson did.

Like many scholars, Lüscher concluded that the poet had a male lover and an intense libidinal life. But Lüscher identified the Master as Samuel Bowles, a family friend and editor of *The Springfield Republican*, not William Smith Clark.

Margulis was skeptical of some aspects of Lüscher's thesis and of his conviction that Bowles was the Master. She believed, nonetheless, that his materials should be published. Dickinson scholar and cognitive linguist Margaret Freeman, who lives in Western Massachusetts, became a consultant on the Lüscher project and prepared materials for presentation to prospective publishers. Ten university publishers turned down the book proposal.

Undaunted, Margulis made Lüscher's case to the Cambridge Forum and its large listenership in February 2011. In her presentation, "Yellow: Decoding Emily Dickinson," Margulis related Lüscher's magnum opus to Jones's evidence that Clark was the Master. (Yellow, in Lüscher's theory, is one of Dickinson's double words and can refer to semen. "Sun" refers to the beginning of ejaculation; "morning" and "east" mean climax; "hat" refers to condom, which were indeed available then; and "sunset" is the male organ undergoing detumescence.) At the forum Margulis presented a half-dozen Dickinson poems as understood in code by Lüscher. She ended by making a heartfelt case for finding an archival home for his papers.

"Anyone who reads the Lüscher materials may be skeptical, annoyed, or disgusted," Margulis later wrote. "But none of us will read many of Dickinson's obscure poems in the same way again."[2]

Margulis edited two essay collections that acknowledge in their titles and in an epigraph Dickinson's injunction to "tell all the truth but tell it slant." The public is not always ready to consider the brilliant insights of women like Dickinson and Margulis, at least not directly. Margulis's generosity toward controversial Dickinson scholarship may have sprung in part from her own experiences outside the mainstream of science. As for her passion for safeguarding archival Dickinson material, Margulis may also have been thinking laterally according to her lifelong habit of leaping across boundaries and synthesizing from a variety of sources and disciplines. Who is to say what evidence, what theory might in time contribute to a new understanding of the nature of things? According to her son and literary collaborator Dorion Sagan, Margulis aimed "to analyze living things and fossils in order to find what the history of life is"—an enormous intellectual and spiritual project, and not so far removed from Dickinson's own.

Terry Y. Allen is an editor and writer who lives in Amherst, Massachusetts.

Jokin' in the Girls' Room

PENNY BOSTON

I can hardly remember a time when I didn't know that there was a Lynn Margulis. I first came across her work when I was in high school, and by the time I hit college I knew that what she was working on was deeply entwined with my own interests. In 1979, still an undergraduate, I went to the Second International Mars Colloquium at Caltech and was amazed to discover that Lynn was in attendance. At the time I didn't realize that published scientists whose names I knew actually had lives and did things like go to meetings.

In that time when women were still almost nonexistent in more scientific fields than we are today, I often found myself the sole patron of the women's restroom at many meetings. At the Mars Colloquium, I was in the girls' room during a break and in walked Lynn—or should I say, "in burst Lynn." She glanced at me and started telling me all about something that I have long ago forgotten. Bowled over, I responded with mmhms and uh-huhs, and made a couple of attempts at humor until we had both washed our hands. In a flurry of last-minute words tossed over her shoulder, Lynn plunged back into the hallway.

I was simultaneously thrilled to have met her, amused at her style, and bemused that I had had a conversation with a real live woman scientist—in the girls' room! Wow! It was a life-changing moment for me, a frame shift in the universe. A talkative, friendly woman who could joke around with a stranger in the ladies' room, she was also a highly accomplished scientist who didn't make a stuffy fuss out of being one. Such an earth mother as she was, I was hooked.

That was only the beginning of knowing Lynn. The following year, 1980, I applied for the new Planetary Biology Intern Program (PBI) that Lynn founded and that still continues. As one of the program's first interns, I worked at Oklahoma State University in Stillwater, Oklahoma, with another great woman of science, Dr. Helen Vishniac. Helen, a mycologist, is the daughter of the twentieth-century paleontologist George Gaylord Simpson and wife of Wolf Vishniac. (Wolf was the developer of the famous Wolf Trap microbiological experiment that was scheduled to fly aboard the *Viking* mission to Mars until it was yanked from the payload more or less at the last minute.[1] Sadly, by the time I worked with Helen on yeasts from the Antarctic Dry Valleys, Wolf had met his fate in an Antarctic crevasse some years before.) The PBI internship marked my transition from student to scientist under Helen's able and demanding but warmhearted tutelage.

Although I ran into Lynn at several more conferences after my PBI stint, I really got to know her during the summer of 1984 at a program called Planetary Biology and Microbial Ecology (PBME) that was housed at San Jose State and NASA Ames Research Center. This field-based research effort was a blend of lectures given by Lynn and other faculty and hands-on research resulting in publication by participants. While the formal part of the program was valuable, the most lasting effect on me was observing Lynn's approach to science, and talking. Just talking. About biology, about what the Gaia hypothesis might and might not mean, about life on other planets, about who was the nicest-looking guy on the grounds crew. I learned to think in depth about what I had been taught in classes and understand where my own opinions diverged from the canon. And where they did not.

The goofiness highlight of that summer—and there were many goofy moments—was a skit at the final social gathering of the group put on by PBME faculty. Lynn portrayed the heroine Lynn Lyngbya (her interpretation of a multicellular cyanobacterium that slips out of its sheath, complete with gliding motion—indescribable!) and commented vociferously on what a bummer it was that *Lyngbya* reproduced asexually.

The next major dose of Lynn that I received was when I was co-convenor of the Chapman Conference on Gaia with another of my great mentors, Dr. Steve Schneider (then of the National Center for Atmospheric Research) in 1988. This meeting brought together people from a plethora of disciplines to consider whether the Gaia hypothesis and related ideas were actually science—that is, whether it was testable in a meaningful way and how one could go about actually testing the hypothesis with various observational and experimental means. We consulted frequently with Lynn as we developed the program, but we also invited people who were not sympathetic to the hypothesis in order to give it a vigorous workout in the broader relevant community.

Amid all this, I was a new mother with an infant daughter; running the meeting, which was held in San Diego in 1988, was my first official work activity after having the baby. Mammalian biology being what it is, my milk dried up, my baby became colicky, and I was on the verge of a meltdown by day two. As an only child of an only child of an only child, I never even babysat in high school, so caring for my baby was definitely a new type of biology project for me. Lynn, mother of four and already a grandmother by then, told me to relax, drink lots of water, and carry on. It would get better. It did.

As my career developed, I discovered that our planet not only has an atmosphere, a surface, and oceans; it also has a large part of its ecosphere tucked below the terrestrial surface in the rock fractures and caves that dot our planetary crust. As I came to know these extraordinary places, and how different they are from the surface, I tried to fit them into my evolutionary worldview. Is the subsurface world Gaian in the sense that Lovelock and Margulis have described? Must a Gaian type of system operate in flamboyant surface abundance, or can it also operate on a planet at a throttled-back level where only a subsurface might be inhabited, somewhere like the subsurface of Mars, or Europa in an ocean under kilometers of ice? I suspect Lynn might still say no, at least she did the last time we talked about it a few years ago. Jim Lovelock might say maybe. I say, "I don't know, but I'm trying to find out."

I often think of Lynn in connection with Beatrix Potter, who as a young woman made an attempt at mycological science and fossil study, only to be rebuffed and ridiculed by the scientific establishment of the time. On the face of it, they were two very different people—Beatrix a shy young Victorian Englishwoman, Lynn a not-so-shy Chicago girl. There was perhaps a contrast in personalities but also a similarity in their abilities to simply say what they saw, an emphasis on the fundamental nature of keen, close observation, even when such observations led to hypotheses and conclusions that were not in vogue at the time. I try to imagine what would happen if, in some alternative universe, they switched places. Beatrix may well still have been too shy to successfully promulgate her views in the world of 1960s American science, while Lynn might have actually been put into an insane asylum in Victorian Britain! I suspect that Lynn would never have been content to retreat to the Lake District countryside to draw the fabulous fuzzy bunnies and fluffy kittens of the Potter stories. I guess we'll never know, but what we can sense in both of these remarkable women is the driving force of observational powers so clear that they would not be talked out of it by those of lesser sight.

Penny Boston is an astrobiologist and founder of the Cave and Karst Studies Program at the New Mexico Institute of Mining and Technology in Socorro, New Mexico, where she is also a professor of earth and environmental science. She is also associate director for academics at the National Cave and Karst Research Institute in Carlsbad, New Mexico. She was an original graduate of NASA's Planetary Biology Internship program. Her TED talk about life on Mars can be seen at: http://www.ted.com/talks/penelope_boston.html.

An Education

EMILY CASE

I first met Lynn in 1997, when I enrolled as a graduate student in her signature course, environmental evolution, at the University of Massachusetts in Amherst. I was working toward a master of education degree, on a yearlong hiatus from teaching. Lynn was then developing a curriculum unit for high school biology about reconstructing past environments using fossil foraminifera and recruited me to help as a summer job. I had been working in a bakery, and Lynn was fond of saying that she found me at a doughnut counter. I was by no means the only student that Lynn rescued from menial labor, and I think this was a particular point of pride with her.

I returned to the Margulis lab every summer thereafter. No other coursework or professional development I've encountered can compete with hanging around Lynn to stay current in my curriculum area, and I felt a great camaraderie with the people I worked with at the lab over the years. You can't imagine a more eclectic group of people; a list of their interests and avocations begins to sound like a joke, something along the lines of "a bouncer, a belly dancer, and a Buddhist walk into a lab . . ."

Here's the punchline: they all walk out with advanced degrees and a few publications under their belts.

One of the difficulties I have faced is explaining to people what Lynn meant to me and what our relationship was. "The scientist I worked for in the summers," sounds awkward and cold, and I once heard Lynn object to this type of description. My fellow Margulis lab employee Jeremy Jorgensen, also a teacher, once introduced himself to a group of

Lynn's students by saying, "I work for Lynn." She made a disapproving sound, so he amended his statement: "I work for science, with Lynn." Of this, she approved.

Lynn was my employer, my mentor, my editor, my coauthor, my friend. But where I'm landing, in my heart, is that she was, first and foremost, my teacher, in the oldest and deepest sense of the word.

Those who know Lynn's work are familiar with her vast expertise, her unparalleled knowledge about the earth and its live beings. To be an apprentice to that was a profound experience.

One of my recent tasks at the Margulis lab was to create a catalog of images that Lynn had collected over her career. Sean Faulkner, who is another student of Lynn's, and I each spent hours peering at micrographs of all kinds scanned from unlabeled or illegible photographic slides. We'd set aside images we could not identify for weeks or months until Lynn was available to help us.

How do you imagine this went?

Most people who hire someone to perform a task like this will have an interest in efficiency. Lynn, presented with the first picture, took a slow breath and said, "Oh, this one is very interesting."

With dozens or even hundreds of images to identify, Lynn would spend an hour explaining just a few of them to Sean or myself. At the time it was pretty frustrating. In retrospect I see that she was always a teacher—the diffusion of knowledge was more important to her than the completion of an image database.

Taking courses with Lynn and working in her lab was a total immersion experience. When I first met Lynn through that fateful environmental evolution course, I was only a few years out of college. I had graduated with honors and a degree in biology from a reputable institution and expected to be well prepared for Lynn's course.

I was not.

Every time I thought we were entering familiar territory—mitosis, say, or genetics, I would quickly learn that almost everything I knew applied only to animals, or sometimes to plants, and thus not at all to the content of this course.

My previous teachers had taught me a great deal; Lynn taught me how much more there is to know: the incredible diversity of metabolic and reproductive strategies among microbes and their evolutionary and planetary implications—this was her field, and I've been coming back for fifteen years to learn more at her side.

Lynn's teaching style was unique. Aaron Haselton, who was a student of Lynn's when I first met him and is now a professor at State University of New York, New Paltz, once offered an apt description of Lynn as a teacher. Aaron was showing a group of new environmental evolution students room 318 of the Morrill Science Center, where the teaching materials were housed. Their faces betrayed a state of bewilderment common to new students of Lynn's, so Aaron paused in his explanations of the teaching resources and began to describe the experience of learning from Lynn.

Some teachers, he explained, present the material sequentially, like this: he traced a neat little outline in the air with his pen. Others emphasize main ideas and connections; he sketched a concept web. Lynn? She dismantles the pen and empties all the ink onto the page at once.

Total immersion. They don't teach this method in education schools, but Lynn was the best teacher I ever had.

I think of the legion of teachers that Lynn taught as her most important educational legacy. In education at all levels, we necessarily simplify concepts to meet the needs of our students, as do our curricula and textbooks. This is an important, difficult, and delicate task, and when it comes to evolution and symbiosis, the standard curricula and textbooks botch the job. In high school textbooks in the United States, the symbiogenetic origins of mitochondria and chloroplasts are nearly always found in a box not in a chapter about evolution, but in a chapter about cells.

We all know what boxes mean. To students they mean that the information is not going to be on the test. To writers and editors, boxes are for neat little facts that don't really fit in with the overall story.

The mathematician John von Neumann said, "In mathematics you don't understand things. You just get used to them." I think the same could be said for science, and probably for any field.

In biology we must take great care to consider just what it is we are getting used to. Is it the astonishing diversity of organisms, life histories, biochemical pathways, metabolic activities, and fossils that we see in nature, or are we simply becoming accustomed to the categories, generalizations, and stories we've created to help our limited minds comprehend life's complexity?

The power of Lynn's scientific ideas comes from her profound knowledge of life itself. Our theories and taxonomies must stand up to ground-truthing at the level of the organism, and Lynn spent a lot of time on the ground, not to mention in the mud. An example of what I mean comes from an exchange between Lynn and Richard Dawkins a few years ago at Oxford in a debate titled "Homage to Darwin":

> Dawkins: If you take the standard story for ordinary animals, you've got a distribution of animals, you've got a promontory, so you end up with two distributions. And then on either side you get different selection pressures, and so one starts to evolve this way, and one starts to evolve that way, and what's wrong with that? It's highly plausible; it's economical; it's parsimonious. Why on earth would you want to drag in symbiogenesis when it's so unparsimonious and uneconomical?
>
> Lynn: Because it's there.

Last summer Lynn urged me to focus my own teaching on the organisms themselves: their bodies, their environments, their activities. Recently, when planning a unit on reproduction and heredity, I recalled Lynn's advice and made the basis of my lesson twelve organisms representing all five kingdoms, with reproductive strategies ranging from binary fission to parthenogenesis. I knew I was doing right by Lynn when one girl looked up at me in the middle of the lesson with wide eyes and said in an awestruck voice, "I, like, seriously did not know any of this before."

I could see her legacy in another student, an extraordinary young man in my class. I can take no credit for his work, except perhaps to say that I did not get in his way—which is, I imagine, what Lynn's best teachers might have said about their work with her.

The week that I learned of Lynn's stroke, my students were giving presentations about organ systems. Almost all of them had chosen human body systems, and a few studied systems in plants. This young man chose fungi. In the midst of good but standard presentations about human digestion and the like was a brilliant, well-researched, captivating discussion of the organization of cells, tissues, and organs in the

Basidiomycota. Later, when we did endangered species projects, among the polar bears, chinchillas, and whales was the rock gnome lichen, from the same student.

To honor Lynn, I am resolved to remember that the truth on the ground is more important—messier but ultimately more beautiful—than the elegance of our explanations. I will breathe life into science education and liberate Lynn's ideas from their boxes. Finally, I will take heart in the knowledge that the next generation still shows wonder at the diversity of life, including the lichens and fungi.

Emily Case teaches life, earth, and environmental science to middle and high school students at Smith Academy, a tiny public school in Hatfield, Massachusetts. From 1997 through 2011, she spent her summers in the Margulis lab, where she assisted with various educational and scientific projects.

There Should Be Other Prizes

DAVID LENSON

In the autumn of the millennial year 2000, a student came into my office in South College to change her major to comparative literature. A routine procedure, except she was coming from management. Unusual conversion—a shift not just in academic concentration but likely in weltanschauung. Brianne Goodspeed was steaming with energy. I signed her up and then spent months enjoying her transmogrification.

Academic advisors don't burn a lot of fat wondering why. Nevertheless, I sensed there must be an esoteric reason for Brianne's breakout from business school—a third location, a waystation between the actuarial and the anarchistic. In time I discovered it: Lynn Margulis's lab. Brianne had been doing something over there.

Now be sure of this: I was not at the time uninterested in science. I hated it. A passionate interest. Science sucked money out of my artsy casserole and turned it into weaponry and spyware. Medicine invented cures worse than diseases. I was responsible for all of culture, but science professors who studied one scummy fungus got paid three times my salary.

Brianne turned into a star in comp lit. And she kept suggesting that I meet Lynn. Brianne was young, thin, and blond, more than alive. Her hands were odd, kind of double-jointed. I looked at her hands. I listened to her boiling mind.

At the time I was cohosting and coproducing a radio program on the campus station, WMUA. It was a spin-off of the *Massachusetts Review*. I was the editor of that.

Just after New Year's 2004, I thought, Okay, I'll invite Lynn on the radio. She accepted, and Brianne came along, beaming conspiratorially in a corner of the studio. Thereafter a good part of my friendship with Lynn was on the radio. When I hadn't seen her for a while and wanted to, I'd just book her again. My coproducer, Roger Fega, and I realized she was the perfect guest: you didn't have to extract words out of her, and she was willing to talk about *anything*.

Between fabulous digressions, she talked about science as she understood it. It dawned on me and Roger that despite the prizes and medals and honors and photos with Bill Clinton, Lynn was actually a dangerous radical. She persuaded me that what I despised wasn't science, but Big Science, which she hated just as much, especially for its subservience to the military. She had a view from the top of that world, and she wasn't guessing.

Here was no Big Scientist but a genuine heir of Aristotle, a lynx-eyed observer of the natural world whose genius was seeing how the denizens of this place, the symbionts, move together to generate life. She said she'd never retire because "they let me go outside." Her last research project was launched because she noticed, literally, a bump on a log.

She also took her responsibilities as Emily Dickinson's next-door neighbor seriously. She dove deep into Emily's work, defended and embodied and expounded it. She translated poetry from Spanish. She knew all about it. Poetry and science were, in her thinking, similar ways of seeing and knowing (though science, naturally, was better).

She believed things, but she understood belief as a process. One time Roger asked her (you could ask a question like this) what happens after death. She replied, "See that little spot of yeast in the corner of your mouth? It takes over everything." Another time we wondered if the human race is doomed to extinction. "Yes," she answered. Roger asked why. "Symbology and genital friction."

It's one thing to be a scientist. It's another to be a full-service intellectual. Lynn won her honors for science, but those of us who knew her realize that there should be other prizes for sheer mentality, for the willingness to consider any problem, political, cultural, poetic, economic, metaphysical, medical, or whatever.

She never took on a problem without problematizing it further. There was never an answer that stopped the investigation. No question was ever settled or dead. She was a process thinker in a field addicted to conclusions. For her, there were no conclusions. There was the trajectory of the mind as far as it could go, and it could always go further. That's why her death is unacceptable. It looks, on the outside, as if the questions were answered, a ribbon tied around the book. It is not so.

Lynn may have died, but what Lynn was all about will never, cannot, die. It's up to the rest of us now: to keep making problems. To keep a keen eye on the world, to see how everything is always changing, to let the mind open into wonderment.

Her ashes were scattered in Puffers Pond, where night after night the cataract falls down in an eternal cascade of questions.

David Lenson, a professor of comparative literature at the University of Massachusetts, is the author of On Drugs *and a saxophone player in the Reprobate Blues Band.*

With Love and Squalor

BETSEY DEXTER DYER

"Don't make me into your role model," she scolded me one day. I was twenty-nine and she was forty-six. We were on a bus in Mexico, returning from a field site. I was teary eyed over some slight or frustration and looking for advice. "You have to make your way on your own!"

"With love and squalor, Lynn," she'd sometimes sign her letters to me, borrowing the phrase from J. D. Salinger.

It was the salutation of a scientist, taking more delight in facing reality than embroidering it.

She was my mentor, my friend, my inspiration. She lived fully, honestly, and without regret. Lynn Margulis would be furious about any banalities about her after death. She was an atheist and a rationalist. Death for her was to be sudden and final. I know because we spent several hours during the summer of 2011 talking about it.

We were at a conference, rooming together for the week in Berlin. Our secondary mission: to seek out our dear old friend Wolfgang Krumbein, who in November of 2010 had had a devastating stroke, rendering him mostly immobile and unable to speak in more than single words. We knew that Wolfi was in a home for seniors somewhere in Berlin. For a good part of the week, we tried to get the address. That gave us plenty to talk about (including late at night, side by side in our comfortable German beds) on the subject of death.

Lynn was vehement that she must go quickly, felled by a single massive stroke or attack and that absolutely no attempt should be made to revive her. She would not live on as just part of herself (as she imagined Wolfi was), confined to wheelchair and bed for as long as the doctors wanted to keep her alive. She graphically described to me that she would demand painkillers, hoard them, and then kill herself with the hoarded supply. It is to the credit of her children that they firmly carried through with Lynn's wishes and signed off on as many papers as the heroic doctors required to be allowed to bring her home to her own bed to slip away, unconscious and unaware.

Although they say she died on November 22 (at home surrounded by family), she really died five days before, on a Thursday after a full day of work, at five in the afternoon, about to mount her bicycle, thinking about dinner, felled by a massive hemorrhagic stroke, never to be Lynn again, her wonderful brain no more.

She left so much not done, and that is entirely a good thing. For she also described the sedentary, retired life of the elderly (venerable) but quite feeble emerita professor (all projects "done") that she resolutely did not want to become. She had, at her death, many projects ranging from nearly finished books to books in the making. She had video projects, including one with some German filmmakers that occupied a good part of our Berlin trip. She had a new organism from Puffers Pond. She was mentoring students at many stages of undergraduate and graduate degrees. There were visitors to her lab both short and long term as well as boarders and diverse other houseguests at home. She had "projects" involving the support, promotion, and celebration of certain renegade scientists operating outside the system and therefore greatly appealing to Lynn. In the queue were invitations to give keynotes, to travel, and to be honored with one or another award or medal. Lately she had been declining those invitations that were going to present unpleasant multi-airport connections, uninteresting destinations, and travel stipends not paid promptly up front.

I am not so sure that Lynn left much unsaid. Certainly that week in Berlin with her gave us plenty of time to account for family, friends, colleagues; projects past, current, and future—all intertwined with delightful gossip, salacious anecdotes, and startling revelations. And so now multiply that times her children and houseguests and faithful lab entourage as well as diverse phone, email, and letter correspondents. She had her say! She made herself understood.

My last meal at her house was in the spring of 2011. I was on my way to pick up my daughter Alice from college. Lunch was the usual abundant, hearty, bubbling affair: sizzling pots and pans of leftover thick soup (new sausages added in), bread, grilled cheese, compotes, and Greek yogurt. And the conversation was just as much a non sequitur as the food: one topic after another often prefaced by, "Do you know so-and-so?" Answer with caution. For with Lynn, there were distinctions. You might have heard of the person (read his paper maybe); maybe you knew *of* them; perhaps you had a conference dinner together years ago, in which case you might be asked to recall details of the conversation! Or perhaps you really knew the person, traveled together, were lab mates, were actual friends.

About three years ago, over breakfast at her house she said she needed someone, right away, to write a chapter on heteroloboseans, a very obscure group of amoebae. No, no I protested. I don't know anything about them! But nobody else does either, and nobody else is capable of pulling together a chapter on them, Lynn replied with finality. Thus I wrote it. And it was within her deadline of three months or so. No regrets except that if I want any more crazy yet serious derailing, presumptuous projects like that I'll have to assign them to myself.

My master's thesis with Lynn on termite symbionts also began with great uncertainty about what I was looking at (or supposed to be looking at), and even worse, it began with a deliberate lie. The lie turned into a fortuitous beginning for a lifelong passion, but it was a lie nonetheless and still haunts me.

It was autumn of 1976, and I was not doing all that well in graduate school at Boston University, mostly my own fault. I had signed up for Lynn Margulis's class in symbiosis. Early in the semester she announced that we were all to open up at least one termite and look at its intestinal contents and to report back our observations the following week. Then she briskly demonstrated what she meant by opening a termite. It is a rather tiny operation, and to this day I have difficulty demonstrating it to more than two or three people at a time. It wasn't Lynn's fault, though. She cheerfully would have opened several more, had I requested it. However, I doubt I leaned in to look closely at her technique, and I even may have walked out of the room (bearing an attitude problem, not yet worked out) before Lynn got her specimen under a scope and was loudly exclaiming (as all termite researchers tend to do) as though seeing the microbes for the first time.

The next week Lynn demanded to know what I had seen with my own termite exploration. It was a small seminar and therefore easy for her to pick out an individual, unfortunately me. I mumbled something like, "Oh, it was pretty good."

Lynn, in front of the other students, proclaimed, "You never even opened up one single termite, did you?" and disdainfully moved on to query the next student. I opened my first termite that day (loved it, of course—words don't suffice), and, well, I guess I am fortunate that with Lynn's forgiving and inclusive personality, she ended up suggesting that I finish off my master's degree with a thesis on termite symbionts.

It happened that I had already gone through a couple of false starts on thesis projects and being ejected from graduate school was a definite possibility. However, thanks to Lynn Margulis and my primary advisor, the parasitologist Stewart Duncan, I completed my study of *Reticulitermes flavipes*, its symbionts having been selectively removed by several different treatments. In subsequent decades I have pursued interests in many different topics of biology, one of the luxuries of being a professor at a small liberal arts college. However, no matter what I am working on, I usually keep on my desk, right in front of my computer monitor, a small jar of termites just for inspiration, like keeping a beautiful rare book on a special stand in the library.

Betsey Dexter Dyer, a professor of biology at Wheaton College, is author of A Field Guide to the Bacteria.

Acknowledgments

I'd like to extend great thanks to the brave, diligent, and loving contributors to this volume, without whom it would not have been possible; to all those behind the scenes who provided their heartfelt words and wishes, their appreciation intellectual and emotional; and to the many wonderful people not included in these pages who gave their kind support to family, colleagues, and friends of Lynn in the aftermath of her sudden death. To Jim MacAllister, Celeste Asikainen, and Mike Dolan, who gave invaluable, multifaceted assistance. To Dave Caron, Christie Lyons, J. Steven Alexander, and Elsa Dorfman for help with the visuals. To Joseph Johnson Cami for translating Jorge Wagensberg's essay from the Spanish, Sean Faulkner for the picture of Lynn in Morocco, Brianne Goodspeed for yeoman work on the manuscript, and to Lois Brynes, Natasha Myers, and Dianne Bilyak for suggestions during composition. This book is for Lynn Margulis's many admirers, fans, friends, colleagues, and supporters around the world.

Selected Works by Lynn Margulis

Books by Lynn Margulis

Origin of Eukaryotic Cells (New Haven: Yale University Press, 1970).

Early Life (Boston: Science Books International, 1982).

Origins of Sex: Three Billion Years of Genetic Recombination (New Haven: Yale University Press, 1986). Coauthor, Dorion Sagan.

Microcosmos: Four Billion Years of Evolution from Our Microbial Ancestors (London: Allen and Unwin, 1987). Coauthor, Dorion Sagan.

Mystery Dance: On the Evolution of Human Sexuality (New York: Summit Books, 1991). Coauthor, Dorion Sagan.

Garden of Microbial Delights: A Practical Guide to the Subvisible World (Dubuque, IA: Kendall Hunt Publishing Company, 1993). Coauthor, Dorion Sagan.

Symbiosis in Cell Evolution: Microbial Communities in the Archean and Proterozoic Eons (New York: W. H. Freeman, 1993).

Slanted Truths: Essays on Gaia, Symbiosis, and Evolution (New York: Copernicus, 1997). Coauthor, Dorion Sagan.

What Is Sex? (New York: Simon & Schuster, 1997). Coauthor, Dorion Sagan.

Five Kingdoms: An Illustrated Guide to the Phyla of Life on Earth (New York: W. H. Freeman, 1998). Coauthor, Karlene V. Schwartz.

Symbiotic Planet: A New Look at Evolution (New York: Basic Books, 1998).

What Is Life? (Berkeley: University of California Press, 2000). Coauthor, Dorion Sagan.

Acquiring Genomes: A Theory of the Origins of Species (New York: Basic Books, 2002). Coauthor, Dorion Sagan.

Dazzle Gradually: Reflections on the Nature of Nature (White River Junction, VT: Chelsea Green Publishing, 2007). Coauthor, Dorion Sagan.

Luminous Fish: Tales of Science and Love (White River Junction, VT: Chelsea Green Publishing, 2007).

Kingdoms and Domains: An Illustrated Guide to the Phyla of Life on Earth (Amsterdam: Academic Press/Elsevier, 2009). Coauthor, Michael J. Chapman.

Books Edited by Lynn Margulis

Symbiosis as a Source of Evolutionary Innovation: Speciation and Morphogenesis (Cambridge: MIT Press, 1991). Edited with René Fester.

Mind, Life and Universe: Conversations with Great Scientists of Our Time (White River Junction, VT: Chelsea Green Publishing, 2007). Edited with Eduardo Punset.

Chimeras and Consciousness: Evolution of the Sensory Self (Cambridge: MIT Press, 2011). Edited with Celeste A. Asikainen and Wolfgang E. Krumbein.

Selected Articles, Interviews, and Videos

PRINT

John Brockman, "Lynn Margulis: Gaia Is a Tough Bitch," in *The Third Culture: Beyond the Scientific Revolution* (New York: Simon & Schuster, 1995).

Lynn Margulis and Emily Case, "The Germs of Life: Our Ancestors Were Bacterial Communities," *Orion*, November/December 2006.

Suzan Mazur, "Lynn Margulis: Intimacy of Strangers and Natural Selection," *Scoop*, 2009, http://www.scoop.co.nz/stories/HL200903/S00194.htm.

Jeanne McDermott, "A Biologist Whose Heresy Redraws Earth's Tree of Life: A Profile of Lynn Margulis," *Smithsonian*, August 1989.

Dick Teresi, "The Discover Interview: Lynn Margulis," *Discover*, April 2011: 66–71.

RADIO

"A Life with Microbes," interview by Paul Evans. BBC Radio 4, *A Life With . . .* , 2009.

VIDEO

Voices of Oxford: Homage to Darwin Debate, May 2009,
Oxford University, England:
Part 1: http://www.voicesfromoxford.com/homagedarwin_part1.html.
Part 2: http://www.voicesfromoxford.com/homagedarwin_part2.html.
Part 3: http://www.voicesfromoxford.com/homagedarwin_part3.html.

Notes

ERUDITION

"Erudition" was first published as "Perspective/Retrospective: Lynn Margulis (1938–2011)" in *Science* 335: 302 (2012) and is used with permission of AAAS.

AS ABOVE, SO BELOW

"As Above, So Below" first appeared in the online magazine *Wild River Review*.

1. John L. Hall, "Spirochete Contributions to the Eukaryotic Genome," *Symbiosis* 54 (2011): 119–129.
2. William Irwin Thompson, ed., *Gaia 2: Emergence: The New Science of Becoming*, (Hudson, NY: Lindisfarne Press, 1991), 50.
3. Ibid., 51.

GAIA IS NOT AN ORGANISM

1. Lynn Margulis, *Symbiotic Planet: A New Look at Evolution* (New York: Basic Books, 1998), 118.
2. "Lynn Margulis," *The Telegraph*, December 13, 2011.
3. James Lovelock, *Homage to Gaia: The Life of an Independent Scientist* (Oxford: Oxford University Press, 2000), 256.
4. J. E. Lovelock, "Gaia as Seen through the Atmosphere," *Atmospheric Environment* 6 (1972): 579.
5. All unpublished correspondence cited in this article was consulted at Lynn Margulis's Environmental Evolution Laboratory, University of Massachusetts at Amherst.
6. Margulis refers to James E. Lovelock and C. E. Giffin, "Planetary Atmospheres: Compositional and Other Changes Associated with the Presence of Life," *Advances in the Astronautical Sciences* 25 (1969): 179–193.
7. Lovelock, *Homage to Gaia*, xi.
8. James E. Lovelock and James P. Lodge Jr., "Oxygen in the Contemporary Atmosphere," *Atmospheric Environment* 6 (1972): 575–578.
9. Lovelock, "Gaia as Seen," 579.
10. Margulis to Lovelock, July 5, 1972.
11. Lovelock to Margulis, October 17, 1972.
12. James E. Lovelock and Lynn Margulis, "Atmospheric Homeostasis by and for the Biosphere: The Gaia Hypothesis," *Tellus* 26 (1974): 2–10.
13. James E. Lovelock and Lynn Margulis, "Homeostatic Tendencies of the Earth's Atmosphere," *Origins of Life* 5 (1974): 99.
14. Lovelock to Margulis, March 2, 1973.
15. Lovelock to Margulis, April 17, 1973.

16. Lynn Margulis and J. E. Lovelock, "Biological Modulation of the Earth's Atmosphere," *Icarus* 21 (1974): 471–489.
17. Lovelock to Margulis, July 21, 1972.

LYNN MARGULIS AND STEPHEN JAY GOULD

1. Stephen Jay Gould, "Foreword" in *Five Kingdoms: An Illustrated Guide to the Phyla of Life on Earth* (New York: W. H. Freeman, 1998), ix.
2. Lynn Margulis and J. E. Cohen, "Combinatorial Generation of Taxonomic Diversity: Implication of Symbiogenesis for the Proterozoic Fossil Record" in *Early Life on Earth*, (New York: Columbia University Press, 1994), 327-333.

TOO FANASTIC FOR POLITE SOCIETY

1. C. Mereschkowsky, "Theorie der zwei Plasmaarten als Grundlage der Symbiogenesis, einer neuen Lehre von der Entstehung der Organismen," *Biologisches Centralblatt* 30 (1910): 277–303, 321–347, 353–367.
2. Ibid.
3. S. Watase "On the Nature of Cell Organization," *Woods Hole Biological Lectures* (1893), 83–103; 86.
4. Boris M. Kozo-Polyansky, *A New Principle of Evolution*, ed. and trans. Victor Fet, ed. Lynn Margulis (Cambridge: Harvard University Press, 2011).
5. I. E. Wallin, "On the Nature of Mitochondria, VII. The Independent Growth of Mitochondria in Culture Media," *American Journal of Anatomy* 33 (1924): 147–173.
6. William Bateson, *Problems of Genetics* (New Haven: Yale University Press, 1913), 990.
7. I. E. Wallin, *Symbionticism and the Origin of Species* (Baltimore: Williams and Wilkins, 1927).
8. Ibid., 147.
9. Félix d'Herelle, *The Bacteriophage and Its Behavior*, trans. George H. Smith, (Baltimore: Williams and Wilkins, 1926), 320.
10. Portier, *Les symbiotes* (Paris: Masson, 1918), 294.
11. Wallin, *Symbionticism*, 8.
12. T. H. Morgan, "Genetics and the Physiology of Development," *American Naturalist* 60 (1926): 489–515; 496.
13. Edmund B. Wilson, *The Cell in Development and Heredity*, 3rd ed. (New York: Macmillan, 1925), 739.
14. Lynn Margulis, "Hans Ris (1914–2004): Genophore, Chromosomes and the Bacterial Origin of Chloroplasts," *International Microbiology* 8 (2005): 145–148.
15. Ibid.
16. Hans Ris and W. Plaut "Ultrastructure of DNA-Containing Areas in the Chloroplasts of *Chlamydomonas*," *Journal of Cell Biology* 13 (1962): 383–391.
17. Sylvan Nass and Margit M. K. Nass, "Intramitochondrial Fibers with DNA Characteristics," *Journal of Cell Biology* 19 (1963): 613–628.
18. Lemuel Rosco Cleveland and A. V. Grimstone, "The Fine Structure of the Flagellate *Mixotricha paradoxa* and Its Associated Micro-organisms" *Proceedings of the Royal Society B*, 159 (1964): 668–686.

19. Lynn Sagan, "On the Origin of Mitosing Cells," *Journal of Theoretical Biology* 14 (1967): 225–274.
20. Ibid., 247.
21. Aharon Gibor, "Inheritance of Cytoplasmic Organelles," in Katherine Brehme Warren, *Formation and Fate of Cell Organelles, Symposia of the International Society for Cell Biology,* vol. 6 (New York: Academic Press, 1967), 305–316.
22. Philip John and F. R. Whatley, "*Paracoccus denitrificans* and the Evolutionary Origin of the Mitochondrion," *Nature* 254 (1975): 495–498. Rudolf A. Raff and Henry R. Mahler, "The Non Symbiotic Origin of Mitochondria," *Science* 177 (1972): 575–582. R. A. Raff and H. R. Mahler, "The Symbiont That Never Was: An Inquiry into the Evolutionary Origin of the Mitochondrion," *Symposia of the Society for Experimental Biology* 29 (1975): 41–92.
23. T. Uzell and C. Spolsky, "Mitochondria and Plastids as Endosymbionts: A Revival of Special Creation?" *American Scientist* 62 (1974): 334–343.
24. R. Stanier, "Some Aspects of the Biology of Cells and Their Possible Evolutionary Significance," in H. P. Charles and B. C. Knight, eds., *Organization and Control in Prokaryotic Cells. Twentieth Symposium of the Society for General Microbiology* (Cambridge: Cambridge University Press, 1970), 1–38; 31.
25. Lynn Margulis, "Symbiotic Theory of the Origin of Eukaryotic Organelles: Criteria for Proof," *Symposia of the Society for Experimental Biology* 29 (1975): 21–38; 21.
26. C. R. Woese and G. E. Fox, "Phylogenetic Structure of the Prokaryote Domain: The Primary Kingdoms," *Proceedings of the National Academy of Sciences USA* 75 (1977): 5088–5090. C. R. Woese, "Endosymbionts and Mitochondrial Origins," *Journal of Molecular Evolution* 10 (1977): 93–96.
27. Michael W. Gray and W. Ford Doolittle, "Has the Endosymbiont Hypothesis Been Proven?" *Microbiological Reviews* 46 (1982): 1–42. See also Michael Gray, "The Endosymbiont Hypothesis Revisited," *International Review of Cytology* 141 (1992): 233–257.
28. John L. Hall and D. J. Luck, "Basal Body-Associated DNA: *In Situ* Studies in *Chlamydomonas reinhardtii,*" *Proceedings of the National Academy of Sciences USA* 92 (1995): 5129–5133. One of the biological lessons from this is that cell structures do not have to have nucleic acids in order to be inherited. See Jan Sapp, "Freewheeling Centrioles," *History and Philosophy of the Life Sciences* 20 (1998): 255–290.
29. H. Hartman and A. Fedorov, "The Origin of the Eukaryotic Cell: A Genomic Investigation," *Proceedings of the Natinoal Academy of Sciences USA* 99: 1420–1425.
30. See Jan Sapp, ed., *Microbial Phylogeny and Evolution: Concepts and Controversies* (New York: Oxford University Press, 2005).
31. John L. Hall, "Spirochete Contributions to the Eukaryotic Genome," *Symbiosis* 54 (2011): 119–129.

32. Sorin Sonea and Maurice Panisset, *A New Bacteriology* (Boston: Jones and Bartlett, 1983); Sorin Sonea and L. G. Mathieu, *Prokaryotology* (Montreal: Les Presses de l'Université de Montréal, 2000).
33. J. Maynard Smith and E. Szathmáry, *The Origins of Life. From the Birth of Life to the Origin of Language* (New York: Oxford University Press, 1999), 107. See also J. Maynard Smith and E. Szathmáry, *The Major Transitions in Evolution* (Oxford: Oxford University Press, 1997), 195.
34. S. Jay Gould, *Wonderful Life: The Burgess Shale and the Nature of History* (London: Hutchison Radius, 1989), 310.

KINGDOMS AND DOMAINS

1. Carl Woese, O. Kandler, and M. Wheelis, "Towards a Natural System of Organisms: Proposal for the Domains Archaea, Bacteria, and Eucarya," *Proceedings of the Natinoal Academy of Sciences USA* 87 (1990): 4576–4579.
2. Motoo Kimura, "Evolutionary Rate at the Molecular Level," *Nature* 217 (1968): 624–626.
3. Robert Harding Whittaker, "New Concepts of Kingdoms or Organisms: Evolutionary Relations Are Better Represented by New Classifications Than by the Traditional Two Kingdoms," *Science* 163 (1969): 150–160.

NEO-DARWINISM

1. Lynn Margulis and Dorion Sagan, *Dazzle Gradually: Reflections on the Nature of Nature* (White River Junction, VT: Chelsea Green Publishing, 2007), 260.

SIPPEWISSETT TIME SLIP

"Sippewissett Time Slip" is adapted from "Extraterrestrial Seas," chapter 7 in *Alien Ocean: Anthropological Voyages in Microbial Seas* (2009) and is used with permission from the University of California Press.

1. Steven J. Dick and James E. Strick, *The Living Universe: NASA and the Development of Astrobiology* (New Brunswick: Rutgers University Press, 2004), 83.
2. Lynn Margulis, "On Syphilis and Nietzsche's Madness: Spirochetes Awake!" *Daedalus* (Fall 2004): 123.
3. Richard Doyle, *Wetwares: Experiments in Postvital Living* (Minneapolis: University of Minnesota Press, 2003), 186.

LYNN MARGULIS ON SPIRITUALITY AND PROCESS PHILOSOPHY

1. This conference, which took place at the Claremont School of Theology in 2004, gave birth to John B. Cobb Jr., ed., *Back to Darwin: A Richer Account of Evolution* (Grand Rapids: William B. Eerdmans, 2008).
2. Alfred North Whitehead, *Science and the Modern World* (1925; New York: Free Press, 1967), 75.
3. Ibid., 103.
4. Ibid., 79.
5. Alfred North Whitehead, *Process and Reality: An Essay in Cosmology*, ed. David Ray Griffin and Donald W. Sherburne (1929; New York: Free Press, 1978), 18.

6. Lynn Margulis, "Gaia and Machines," in *Back to Darwin*, 167–75; 172.
7. Dick Teresi, "Lynn Margulis Says She's Not Controversial, She's Right," *Discover*, April 2011 (http://discovermagazine.com/2011/apr/16-interview-lynn-margulis-not-controversial-right/article_view?b_start:int=0&-C).
8. John Brockman, "Lynn Margulis: Gaia Is a Tough Bitch," chap. 7 in *The Third Culture: Beyond the Scientific Revolution* (New York: Simon & Schuster, 1995); also online (http://www.edge.org/documents/ThirdCulture/n-Ch.7.html).
9. David Ray Griffin, *Unsnarling the World-Knot: Consciousness, Freedom, and the Mind-Body Problem* (Berkeley: University of California Press, 1998; reprint, Eugene: Wipf and Stock, 2008).
10. John B. Cobb Jr., "Organisms as Agents in Evolution," chap. 15 in *Back to Darwin.*
11. Lynn Margulis and Dorion Sagan, *What Is Life?* (New York: Simon & Schuster, 1995), 220.
12. Lynn Margulis, "Quite Well," John Templeton Foundation, "Does Evolution Explain Human Nature?" (http://www.templeton.org/evolution/).
13. Margulis, "Gaia and Machines," 173.
14. Lynn Margulis and Dorion Sagan, "The Role of Symbiogenesis in Evolution," *Back to Darwin*, 176–184; 182.
15. Whitehead, *Science and the Modern World*, 110.
16. Charles Hartshorne, "The Compound Individual," in *Philosophical Essays for Alfred North Whitehead*, ed. Otis Lee (New York: Longmans, Green & Co., 1936) reprinted in Hartshorne, *Whitehead's Philosophy: Selected Essays, 1936–1970* (Lincoln: University of Nebraska Press, 1972).
17. Ibid.
18. Brockman, "Lynn Margulis: Gaia Is a Tough Bitch."
19. Ibid.
20. Stated by Dawkins in *The Third Culture* (New York: Simon & Schuster, 1995), 144.
21. Margulis, "Serial Endosymbiotic Theory (SET) and Composite Individuality: Transition from Bacterial to Eukaryotic Genomes," Microbiology Today 31 (2004): 172–174 (http://www.sgm.ac.uk/pubs/micro_today/pdf/110406.pdf).
22. Lynn Margulis and Dorion Sagan, *Acquiring Genomes: A Theory of the Origins of Species* (New York: Basic Books, 2002), 97.
23. Margulis, "Symbiogenesis and Symbionticism," in *Symbiosis as a Source of Evolutionary Innovation*, ed. Lynn Margulis and René Fester (Cambridge: MIT Press, 1991), 1–12; 10.
24. Brockman, "Lynn Margulis: Gaia Is a Tough Bitch."
25. These quoted statements are in Margulis's account of the views of Harvard University's Heinrich ("Dick") Holland, in "Gaia and Machines" (in *Back to Darwin*), 172.
26. Whitehead, *Science and the Modern World*, 112.
27. Alfred North Whitehead, "From Force to Persuasion," chap. 5 in *Adventures of Ideas* (1933; New York: Free Press, 1967).
28. Margulis, "Quite Well."

29. Lynn Margulis and Dorion Sagan, "Marvellous Microbes," *Resurgence* 206 (2001): 10–12.
30. Lynn Margulis and Dorion Sagan, *Microcosmos: Four Billion Years of Microbial Evolution* (Berkeley: University of California Press, 1997), 15–16.
31. These statements by Dennett and Williams are in Brockman's *The Third Culture*, at the end of chapter 7 (after Margulis's "Gaia Is a Tough Bitch").
32. Eric Goldscheider, "Evolution Revolution," *On Wisconsin Magazine* (Fall 2009) (http://onwisconsin.uwalumni.com/features/evolution-revolution/).
33. David Ray Griffin, The New Pearl Harbor Revisited: 9/11, the Cover-Up, and the Exposé (Northampton: Olive Branch Press, 2008); John B. Cobb Jr., "Truth, 'Faith,' and 9/11," in Matthew J. Morgan, ed., The Impact of 9/11 on Religion and Philosophy: The Day That Changed Everything? (New York: Palgrave Macmillan, 2009), 151–168.
34. David Ray Griffin, *The New Pearl Harbor: Disturbing Questions about the Bush Administration and 9/11* (Northampton: Olive Branch Press, 2004).
35. Patriots Question 9/11 (http://patriotsquestion911.com/professors.html #Margulis).
36. Lynn Margulis, "Two Hit, Three Down—The Biggest Lie," *Rock Creek Free Press*, January 24, 2010 (http://rockcreekfreepress.tumblr.com/post/353434420/two-hit-three-down-the-biggest-lie).

A FEROCIOUS INTELLIGENCE

1. David Abram, "The Perceptual Implications of Gaia," *The Ecologist* (1985), 15, 3.
2. David Abram, "The Mechanical and the Organic: On the Impact of Metaphor in Science," in *Scientists on Gaia*, ed. Stephen Schneider and Penelope Boston (Cambridge: MIT Press, 1991).
3. Ibid.
4. David Abram, *The Invisibles: Selected Essays*, forthcoming from Pantheon Books. Originally published as "The Invisibles," in *Parabola* 31, 1 (Spring 2006).

FISHERMEN IN THE MAELSTROM

1. Norbert Elias, *Involvement and Detachment* (London: Blackwell, 1987).
2. Robert Poole, *Earthrise: How Man First Saw the Earth* (New Haven: Yale University Press, 2008).
3. Tim Lenton and Andrew Watson, *Revolutions That Made the Earth* (Oxford: Oxford University Press, 2011).
4. Peter Westbroek, *Life as a Geological Force* (New York: Norton, 1991).
5. Lynn Margulis, *Symbiotic Planet: A New Look at Evolution* (New York, Basic Books, 1998).
6. Edgar Morin, *La méthode* (Paris: Seuil, 1977–2004).
7. Westbroek, *Life as a Geological Force.*
8. Norbert Elias, *The Civilizing Process* (London: Blackwell, 2000).
9. Peter Westbroek, "Civilizing Earth," *Human Figurations*, January 2012.
10. Kevin Kelly, *What Technology Wants* (New York: Viking, 2011).

11. Elias, *Involvement and Detachment.*
12. Sorin Sonea and Maurice Panisset, *A New Bacteriology* (Boston: Jones and Bartlett, 1983).
13. Frank Ryan, *The Mystery of Metamorphosis: A Scientific Detective Story* (White River Junction, VT: Chelsea Green Publishing, 2011).
14. David Christian, *Maps of Time: An Introduction to Big History* (Berkeley: University of California Press, 2004).
15. Fred Spier, *Big History and the Future of Humanity* (Malden: Wiley-Blackwell, 2010).

GAIADELIC

1. Lynn Sagan, "An Open Letter to Mr. Joe K. Adams," *Psychedelic Review* 1, 3 (1964): 354.
2. Ibid.
3. Joe K. Adams, *Psychedelic Review* 1, 2 (1963): 125.
4. Ibid., 128.
5. Sagan, 354.
6. Ibid.
7. Ibid., 355.
8. Ibid.
9. Hofmann, http://www.psychedelic-library.org/child1.htm.
10. Roland R. Griffiths, William A. Richards, Matthew W. Johnson, Una D. McCann, and Robert Jesse, "Mystical-Type Experiences Occasioned by Psilocybin Mediate the Attribution of Personal Meaning and Spiritual Significance 14 Months Later," *Journal of Psychopharmacology* 22, 6: (2008) 621–632.
11. Lynn Sagan, "On the Origin of Mitosing Cells," *Journal of Theoretical Biology* 14 (1967): 225.
12. Ibid., 228.
13. Lynn Margulis and Dorion Sagan, *Acquiring Genomes: A Theory of the Origins of Species* (New York: Basic Books, 2002), 48.
14. James E. Lovelock and Lynn Margulis, "Atmospheric Homeostasis by and for the Biosphere: The Gaia Hypothesis," *Tellus* 26 (1974): 1–2; 3.

TWO HIT, THREE DOWN

"Two Hit, Three Down" first appeared in the *Rock Creek Free Press*, January 24, 2010.

1. John B. Cobb Jr., ed., *Back to Darwin: A Richer Account of Evolution* (Grand Rapids: William B. Eerdmans, 2008).
2. David Ray Griffin, *The New Pearl Harbor: Disturbing Questions about the Bush Administration and 9/11* (Northampton: Olive Branch Press, 2004).
3. David Ray Griffin, *Christian Faith and the Truth behind 9/11: A Call to Reflection and Action* (Louisville, KY: Westminster John Knox Press, 2006).
4. David Ray Griffin, *The Mysterious Collapse of World Trade Center 7: Why the Final Official Report about 9/11 Is Unscientific and False* (New York: Olive Branch Press, 2009).

NO SUBJECT TOO SACRED

1. "The Last Word with John Wilson," BBC Radio 4, December 16, 2011, http://www.bbc.co.uk/iplayer/episode/b0184w5r/Last_Word_Christopher_Hitchens_Lynn_Margulis_George_Whitman_and_Jerry_Robinson/.
2. "Lynn Margulis," *The Telegraph*, December 13, 2011, http://www.telegraph.co.uk/news/obituaries/science-obituaries/8954456/Lynn-Margulis.html.
3. "The Last Word with John Wilson."
4. Ibid.
5. Richard Evans Schultes, "Antiquity of the Use of New World Hallucinogens," *The Heffter Review of Psychedelic Research* 1 (1998).
6. Dick Teresi, "Lynn Margulis," *Discover*, April 2011: 66–71.
7. "The Last Word with John Wilson."
8. Ramon Gilsanz and Willa Ng, "Single Point of Failure: How the Loss of One Column May Have Led to the Collapse of WTC 7," *Structure*, November 2007, http://www.structuremag.org/article.aspx?articleID=284.
9. Niels H. Harrit, et al. "Active Thermitic Material Discovered in Dust from the 9/11 Word Trade Center Catastrophe," *The Open Chemical Physics Journal* 2 (2009): 7–31.
10. Architects and Engineers for 9/11 Truth, "9/11: Explosive Evidence–Experts Speak Out," http://www.youtube.com/watch?v=g-GFBEX5bjY.

NEXT TO EMILY DICKINSON

1. Lynn Margulis, "For Nature Is a Stranger Yet" (unpublished manuscript, September 11, 2011), courtesy of Ruth Owen Jones.
2. Ibid.

JOKIN' IN THE GIRLS' ROOM

1. E. C. Ezell and L. N. Ezell, "On Mars: Exploration of the Red Planet, 1958–1978," Washington, DC: Scientific and Technical Information Branch, National Aeronautics and Space Administration, 1984. http://history.nasa.gov/SP-4212/ch7-4.html.

Index

NOTE: Page numbers in *italics* refer to the photographs on the color insert pages. LM refers to Lynn Margulis.

the politics and practice of sustainable living

CHELSEA GREEN PUBLISHING

Chelsea Green Publishing sees books as tools for effecting cultural change and seeks to empower citizens to participate in reclaiming our global commons and become its impassioned stewards. If you enjoyed reading *Lynn Margulis*, please consider these other great books related to Sciencewriters and Science.

DAZZLE GRADUALLY
Reflections on the Nature of Nature
LYNN MARGULIS AND DORION SAGAN
9781933392318
Paperback • $25.00

MIND, LIFE, AND UNIVERSE
Conversations with Great Scientists of Our Time
EDITED BY LYNN MARGULIS AND EDUARDO PUNSET
9781933392431
Paperback • $21.95

LUMINOUS FISH
Tales of Science and Love
LYNN MARGULIS
9781933392332
Hardcover • $21.95

DEATH & SEX
TYLER VOLK AND DORION SAGAN
9781603581431
Hardcover • $25.00

For more information or to request a catalog,
visit **www.chelseagreen.com** or
call toll-free **(802) 295-6300**.